# SUCCESS AT AQA PHYSICS B AS

Ken Price & Gerard Kelly

**OXFORD**
UNIVERSITY PRESS

# OXFORD
UNIVERSITY PRESS

Great Clarendon Street, Oxford OX2 6DP

Oxford University Press is a department of the University of Oxford.
It furthers the University's objective of excellence in research, scholarship,
and education by publishing worldwide in

Oxford  New York

Athens  Auckland  Bangkok  Bogotá  Buenos Aires  Cape Town
Chennai  Dar es Salaam  Delhi  Florence  Hong Kong  Istanbul
Karachi  Kolkata  Kuala Lumpur  Madrid  Melbourne  Mexico City  Mumbai
Nairobi  Paris  São Paulo  Shanghai  Singapore  Taipei  Tokyo  Toronto  Warsaw

with associated companies in Berlin  Ibadan

Oxford is a registered trade mark of Oxford University Press
in the UK and in certain other countries

© Ken Price and Gerard Kelly 2000

The moral rights of the authors have been asserted

Database right Oxford University Press (maker)

First published 2000
Reprinted 2000, 2001

All rights reserved. No part of this publication may be reproduced,
stored in a retrieval system, or transmitted, in any form or by any means,
without the prior permission in writing of Oxford University Press,
or as expressly permitted by law, or under terms agreed with the appropriate
reprographics rights organization. Enquiries concerning reproduction
outside the scope of the above should be sent to the Rights Department,
Oxford University Press, at the address above

You must not circulate this book in any other binding or cover
and you must impose this same condition on any acquirer

British Library Cataloguing in Publication Data
Data available

ISBN 0-19-914800-7

Typeset in ITC Charter
by Cambridge Publishing Management
Printed in Italy
by G. Canale & C. S.p.A., Borgaro T.se, Turin

# Contents

## Introduction — 5
About this Book — 5
Practical Work — 5
The AS Examination — 5
Units 1 and 2 — 6
Unit 3 — 6

## ■ Revising for Examinations — 7

## ■ Key Words in Examination Papers — 8

## ■ Formulae — 9

## ■ Unit Three Experimental Skills — 11
Planning — 11
Implementing — 11
Analysing — 11
Evaluating — 14
Uncertainties — 14

## Unit 1 — 17

### ■ Scalars and Vectors — 17
Scalar Quantities — 17
Vector Quantities — 17
Adding Forces at Right Angles — 17
Scale Drawing — 17
Calculation — 18
Addition of Other Vector Quantities — 18
Resolving Vector Quantities — 18
Resolution of Velocities — 19
Resolving Forces in Other Directions — 19
Calculating Moments — 20

### ■ Kinematics — 23
Velocity and Speed — 23
Distance–time Graphs — 23
Acceleration — 23
Speed–time Graphs — 24
Finding Distance Travelled from a Speed–time Graph — 24
Finding the Acceleration Due to Gravity, $g$ — 25
Equations of Motion — 25
Derivation of the Equations of Motion — 25
Using the Equations of Motion — 26
Non-uniform Acceleration — 26
Bodies Moving Horizontally and Vertically at the Same Time — 27

### ■ Springs, Energy and Work — 28
Energy Stored in a Stretched Spring — 28
Energy Released from a Stretched Spring — 28
Electrical Energy Sources — 29
Choosing How to Produce Electricity — 30
Original Sources of Energy — 30
Distribution of Electricity — 30
Advantages and Disadvantages — 31
Work — 32
Gravitational Potential Energy — 32
Kinetic Energy — 33

### ■ Forces and Fluids — 34
Lift Forces — 34
Drag — 35
Air Resistance — 36
Maximum Speed of Vehicles — 36

### ■ Newton's Laws of Motion — 37
Equilibrium — 37
Weight and Mass — 37
Acceleration Due to Gravity — 37

### ■ Mechanical Oscillations — 39
Finding the Acceleration Due to Gravity, $g$ — 39
Mass/Spring Oscillators — 40
Damping of Oscillations — 40
Using Data Logging to Investigate Oscillations — 40

### ■ Electricity — 42
Current ($I$) — 42
The Ampere (amp) — 42
The Relationship between Current and Charge — 42
Movement of Electrons and Lattice Ions — 42
Current, Charge and Drift Velocity — 42
Conduction in Liquids and Gases — 43
Resistance ($R$) — 43
Resistance in Metals — 44
Variation of Resistance with Temperature — 44
The Heating Effect of Currents — 44
Factors Affecting the Resistance of Wires — 44
Resistance of Other Components — 45
Superconductivity — 46
Voltage and Electromotive Force (emf) — 47
Potential Difference — 47
Potential Differences in Series Circuits — 47
Potential Difference across Components in Parallel — 47
Current at Junctions in a Circuit — 48
Resistors in Series — 48
Resistors in Parallel — 48
Emf and Internal Resistance — 48
Significance of Internal Resistance — 49
Energy and Power in Circuits — 49
Power Loss in Electricity Cables — 50
Alternating Current for Mains Power — 50
Potential Dividers — 51
Using Potential Dividers — 51
Potentiometers — 51

## Unit 2 — 53

### ■ Waves — 53
What is a Wave? — 53
Making Waves — 53
Periodic Motion — 53
Definitions — 53
Relationship between $f$ and $T$ — 54
Oscillations of Particles in Waves — 54
Snapshot of a Wave — 54
Travelling Waves — 54
What is a Wave Front? — 54
Relationship between Velocity, Wavelength and Frequency — 55
Phase Difference — 55
Complex Waves — 56
Transverse Waves — 56

| | |
|---|---|
| Longitudinal Waves | 57 |
| Sound Waves | 57 |
| What are Light Waves and Radio Waves? | 57 |
| Wave Speeds | 57 |
| Radar | 57 |
| Ultrasound | 57 |
| Measuring the Speed of Sound | 58 |
| Properties of Waves | 59 |
| Reflection | 60 |
| Refraction | 60 |
| Refraction in Communications | 61 |
| Absorption | 61 |
| Polarization | 62 |
| Fibre Optic Cables | 62 |

### Diffraction and Interference — 64

| | |
|---|---|
| What is Diffraction? | 64 |
| Diffraction and Radio Transmission | 65 |
| Energy from Aerials | 66 |
| Loudspeaker Design | 66 |
| Resolution | 66 |
| Superposition of Waves | 68 |
| Synthesizing Complex Sounds | 68 |
| Interference | 68 |
| Path Difference | 70 |
| Two-source Interference | 70 |
| Monochromatic Light | 71 |
| Optical Interference: Young's Experiment | 71 |
| Interference using Reflections | 72 |
| Interference between Waves in Communication | 73 |
| Diffraction Gratings | 75 |

### Stationary Waves — 78

| | |
|---|---|
| Production of a Stationary Wave | 78 |
| Nodes and Antinodes | 78 |
| How do the Different Parts of the Medium Move? | 78 |
| Stationary Waves on a Stretched String | 78 |
| Fundamental Frequency of a Stretched String | 79 |
| Overtones or Higher Harmonics | 79 |
| Frequencies of Modes of Vibration | 79 |
| How does a Stretched String Vibrate? | 79 |
| Investigating the Factors that Affect the Frequency | 80 |
| Testing $f \propto 1/l$ | 80 |
| Phenomena that Involve Stationary Transverse Waves | 80 |
| Looking Forward | 80 |

### Electromagnetic Waves — 82

| | |
|---|---|
| Speed of Electromagnetic Radiation | 82 |
| Photons | 82 |
| The Electromagnetic Spectrum | 82 |
| Emission Spectra | 84 |
| Absorption Spectra | 84 |

### Doppler Effect — 86

| | |
|---|---|
| Uses of the Doppler Effect | 86 |
| The Hubble Law | 87 |
| The Parsec | 87 |
| The Age of the Universe | 88 |

### Atomic Structure — 89

| | |
|---|---|
| Proton and Nucleon Numbers | 89 |
| Nuclear Nomenclature | 89 |
| Nuclides and Isotopes | 89 |
| Graphs of Neutrons against Protons | 90 |
| Radioactivity | 91 |
| Alpha Emission | 91 |
| Beta⁻ Emission | 92 |
| Positron Emission | 92 |
| Gamma Radiation | 92 |
| Evidence for the Existence of Neutrinos | 93 |
| Ionization | 94 |
| Typical Ranges of Radiation | 94 |
| Activity | 95 |
| Count Rate | 95 |
| Background Radiation | 95 |
| Randomness of Radioactive Decay | 96 |
| The Decay Equation | 96 |
| Half-life | 96 |
| Exponential Changes | 97 |
| Practical Graphs | 97 |
| Half-life of Long Half-life Sources | 98 |
| Radiocarbon Dating | 98 |
| Energy Sources | 98 |
| Detecting Radiation | 100 |
| Conducting Experiments with Radiation | 100 |
| Inverse Square Law for Gamma Radiation | 101 |
| Physiological Effects of Radiation | 102 |
| Effect of Radiation on Materials | 102 |
| Some Industrial Uses of Radioactive Sources | 102 |
| Some Medical Uses of Radioactive Sources | 103 |

### Probing the Nucleus — 105

| | |
|---|---|
| The Existence of the Nucleus | 105 |
| Protons in the Nucleus | 105 |
| Cloud Chambers | 106 |
| The Discovery of Neutrons | 106 |
| Bubble Chambers | 106 |
| Confirmation of Particle Types in Cloud and Bubble Chamber Pictures | 107 |
| 'Seeing' the Particles | 107 |
| Classification of Particles | 107 |
| Particle Properties | 107 |
| Leptons | 107 |
| Muons and Tau Leptons | 108 |
| Mesons | 108 |
| Hadrons | 108 |
| Quarks | 108 |
| Baryon Number | 108 |
| Stability | 108 |

### Nature of Information — 110

| | |
|---|---|
| Advantages of Data Capture | 110 |
| Analogue Instruments | 110 |
| Computer Monitoring and Remote Sensing | 110 |
| Sensors | 111 |
| Current Measurement | 111 |
| Digital Data | 112 |
| Sampling Data | 113 |
| Transmitting Data | 113 |
| Thermal Energy and Miniaturization | 114 |
| Information using X-rays and Ultrasound – Comparison of Techniques | 114 |
| Communication Systems | 115 |
| Audio Information and Bandwidth | 116 |
| Analogue Information | 116 |
| More Complex Sounds: Speech Transmission | 117 |
| Base Bandwidth | 117 |
| Modulation | 118 |
| Channel Bandwidth for Audio Information | 118 |
| Video Information | 118 |
| Transmitting more Channels | 118 |
| Digitizing Information | 119 |
| Sampling Rates | 119 |
| Converting Analogue Voltages to Digital Signals | 120 |
| Time-division Multiplexing | 121 |

### Answers — 122

### Index — 131

# Introduction

## About this book

In this book you will find:
- information about the examination papers you will take
- advice on how to tackle question papers effectively
- the definitions and facts you must learn
- the principles you should know, understand and be able to use in problems
- examples of the relevance of the principles in everyday life, particularly in applications relating to information gathering and processing
- examples for you to try to check your learning and your ability to apply ideas.

The book covers three units. The practical skills you will need for Unit 3 are dealt with first in this chapter. Most topics in Units 1 and 2 are covered in sequence. Ideas that support the theme of Information and Communication are at the end of the book.

For Unit 1 you will need to have studied the work on sensors, the way they are used to gather information from experiments and their conversion into a form suitable for transmission.

For Unit 2 you will also need to understand these same principles and have a basic understanding of how information is transmitted.

## Practical Work

There is more to the study of physics than ideas in a book. A good scientist needs good practical skills as well as a sound knowledge base, so it is important that you gain as much practical experience as you can. Unit 3 tests how well you have developed your practical skills.

Through your practical assignments you will:
- improve your understanding of the principles in particular topics
- develop organizational skills, problem-solving skills, skills using equipment and analytical skills that will be of benefit whatever you decide to do after your AS course
- be able to gather evidence for your key skills portfolio.

It is therefore important to make the best use of your practical time and use every opportunity to practise writing down and to discuss with others what you have found.

Use this book as the framework for your studies but find time to explore the ideas further. Use a variety of resources such as textbooks, CD-roms and the internet. You will find that one resource may put the ideas in a way that explains them more clearly than another.

## The AS Examination

The AS examination consists of **three units**.

**Table I.1**

|  | Unit 1 | Unit 2 | Unit 3 |
| --- | --- | --- | --- |
| Content | Unit 1 | Unit 2 | Practical skills |
| Time allowed | $1\frac{1}{2}$ h | $1\frac{1}{2}$ h | 2 h |
| Marks in paper | 75 | 75 | 78 |
| % of AS | 35 | 35 | 30 |
| % of AL | 17.5 | 17.5 | 15 |

There is no choice in any of the examination papers.

## Units 1 and 2

These consist of short-answer questions and structured questions based on the unit content. Examples of questions are included within topics and at the end of each section.

The following are some points to note when taking an examination:
- Read the question carefully. Make sure you understand exactly what is required. Use the key words (page 8) to guide you.
- If you find that you are unable to do a part of a question do not give up. The next part may be easier and may provide a clue to what you might have done in the part you found difficult.
- Note the number of marks per question as a guide to the depth of response needed.
- Underline or note the key words that tell you what is required.
- Underline or note data as you read the question.
- Structure your answers carefully.
- Show all steps in calculations. Include the equations you use and show the substitution of data. Remember to work in SI units.
- Make sure your answers are to suitable significant figures (usually two or three).
- Include a unit unless the answer is a pure number or a ratio.
- Consider whether the magnitude of a numerical answer is reasonable for the context. If it is not, check your working.
- Draw diagrams and graphs carefully.
- Read data from graphs carefully; note scales and prefixes on axes.
- Keep your eye on the clock but do not panic.
- If you have time at the end, use it. Check that your descriptions and explanations make sense. Consider whether there is anything you could add to an explanation or description. Repeat calculations to ensure that you have not made a mistake.

## Unit 3

The practical examination tests your practical skill at the end of the AS course.

You will need to use your knowledge of physics to answer some questions.

The examination consists of three compulsory exercises:
- two short questions each taking 30 minutes
- one long question, which will take 1 hour.

You will be tested on your abilities in four aspects of practical work:
- planning experiments
- implementing or carrying out experiments
- analysing data
- evaluating methods.

### Short questions

In one of the short questions you will be asked to make some observations and then use these to **plan** an experiment. You will not have to carry out the experiment.

In the other short question you will obtain some data, **analyse** the data to draw conclusions, estimate uncertainties and **evaluate** the task, suggesting improvements.

### Long question

The long question will ask you to carry out an experiment. You will be provided with general instructions telling you what you have to do but it is up to you to ensure that the experiment is carried out efficiently using the relevant skills that you have learned during the course. Below are some of the important things you need to remember.

### Carrying out experiments during the course

During the course you will carry out experiments to show how theories work in practice and to help you understand the theories.

All the skills tested in the examination are those you should acquire in your day-to-day activities during the course.

Remember to do the following in your course work:
- keep good records of everything that you do to practise the skills that are tested
- make sure that you have hands-on experience of all the apparatus that is used
- do not leave all the practical work to others when working as a group.

Some of the skills tested are those that are common to all experiments, such as tabulation and graph drawing. Others are specific to a type of experiment, such as procedures to determine accurate values for a period or a resistance. Good records will ensure that you can revise these in preparation for the practical examination.

# Revising for Examinations

There is no one method of revising that works for everyone. It is therefore important to find the approach that suits you best. The following rules may serve as general guidelines.

**Give yourself plenty of time:** Do not leave everything until the last minute. This reduces your chances of success and you could start to panic, which will reduce your concentration. Few people can revise everything the night before and still do well in an examination the next day.

**Plan your revision timetable:** Plan your revision timetable some weeks before the examination and follow it. Do not be side-tracked. Allow time for unforeseen problems that could arise, such as illness.

**Relax:** Concentrated revision is very hard work. It is as important to give yourself time to relax as it is to work. Build some leisure time into your revision timetable.

**Give yourself a break:** Work for about an hour then take a break for 15 to 20 minutes. Then go back to another productive revision period.

**Find a quiet corner:** Find the conditions in which you can revise most efficiently. Many people think they can revise in a noisy, busy atmosphere but most cannot. Any distraction lowers concentration. Revising in front of a television does not generally work.

**Keep track:** Use a checklist and the specification to keep track of your progress. Mark off topics you have revised and feel confident with. Concentrate your revision on things you are less happy with.

**Make your own notes and use colours:** Revision is often more effective when you do something active rather than simply reading. The key ideas are in this book. Reading other books with different approaches might help you understand better.

**Concentrate on understanding:** Memorizing definitions, facts, etc. is important but try to understand how the ideas are applied.

**Practise answering questions:** As you finish each topic, try answering the questions within and at the end of each section or questions from past papers. Even when you have done them before, the act of remembering how they are done will help you learn.

# Key Words in Examination Papers

How you respond to a question can be helped by studying the following, which are the more common key words used in examination questions.

### Name

The answer is usually a technical term or its equivalent, consisting of one or two words.

### State

This requires a statement of a fact, a procedure, a principle or an equation without elaboration. Where further comment is needed the question will ask you to 'State and explain'.

### Explain

The important thing to note here is that a reason or explanation must be given, not just a description.

### List

You need to write down a number of points (each may be a single word) with no elaboration.

### Define/What is meant by?

'Define' requires you to give a precise meaning of a particular term. 'What is meant by' is used to emphasize that a formal definition is not required.

### Outline

You need to give a brief summary of the main points. The mark allocation is a good guide to the detail required.

### Describe

The answer is a description of an effect, experiment or possibly a graph shape. No explanations are required.

### Describe how you would

This is usually used in the context of experiment design. It requires you to say how an experiment could be done by you as a student working in your AS laboratory.

### Suggest

This is used when it is not possible to give the answer directly from the facts that form part of the subject material in the specification. The answer may be based more on general understanding than on recall of learned material.

### Calculate/Determine a value for/Determine the magnitude of/Find

A numerical answer is required, usually using data given in the question. Remember to give your answer to a suitable number of significant figures and give a unit.

### Justify

This is similar to 'Explain'. You will have made a statement and now have to provide a reason for giving that statement.

### Draw/Make a drawing/Sketch a diagram

You simply draw a diagram. You will usually be asked to label it. It is sensible to provide labelling even when this is not asked for.

### Sketch a graph

You need to draw the general shape of the graph on labelled axes. You should include enough quantitative detail to show relevant intercepts and/or whether the graph is exponential or some inverse function, for example.

### Plot

The answer will be an accurate plot of a graph on graph paper. Often there is also a question asking you to determine some quantity from the graph or to explain the shape of the graph.

### Estimate

You may need to use your knowledge and/or your experience to deduce the magnitude of some quantities to arrive at the order of magnitude for some other quantity defined in the question.

### Discuss

This will require an extended response in which you demonstrate your knowledge and understanding of a given topic.

### Show that

You will have been given either a set of data and a final value (which may be approximate) or an algebraic equation. You need to show clearly all the basic equations that you use and all the steps that lead to the to the final answer. Remember to quote the result of your working and then relate this to the value given in the question.

# Formulae

## Formulae

In your revision remember to:
- learn the formulae that are not on your formula sheet
- make sure that you know what the symbols represent in learned formulae and those on the formula sheet
- make sure you take account of prefixes (Table I.2) and learn how to convert them to standard form.

You will be provided with the formula sheet shown on page 10. (Note that occasionally there may be modifications to the formula sheet shown on page 10. Your teacher should be able to tell you if there have been changes.)

**Table I.2**

| Prefix | Meaning | Multiplying factor |
|---|---|---|
| G | giga | $10^9$ |
| M | mega | $10^6$ |
| k | kilo | $10^3$ |
| c | centi | $10^{-2}$ |
| m | milli | $10^{-3}$ |
| μ | micro | $10^{-6}$ |
| n | nano | $10^{-9}$ |
| p | pico | $10^{-12}$ |

## Formulae to remember

You will need to be able to recall and use the following mechanics formulae:

$$v = \frac{s}{t} \qquad F = ma \qquad \rho = \frac{m}{V}$$

$$W = Fs \qquad P = \frac{W}{t} \qquad \text{weight} = mg$$

$$\text{kinetic energy} = \tfrac{1}{2}mv^2 \qquad \Delta(\text{potential energy}) = mg\Delta h$$

Table I.3 shows what each symbol represents.

**Table I.3**

| Symbol | Quantity |
|---|---|
| $s$ | Displacement |
| $v$ | Speed or magnitude of velocity |
| $t$ | Time |
| $F$ | Force |
| $m$ | Mass |
| $a$ | Acceleration |
| $\rho$ | Density |
| $V$ | Volume |
| $P$ | Power |
| $W$ | Work done or energy transferred |
| $g$ | Acceleration of free fall |
| $h$ | Height |
| $\Delta$ | Change in (e.g. $\Delta h$ = change in height) |

You will need to be able to recall and use the following electricity formulae:

$$\Delta Q = I\,\Delta t \qquad V = IR \qquad P = VI$$

$$\text{potential difference} = \frac{W}{Q} \qquad R = \frac{\rho l}{A}$$

$$\text{energy} = VIt$$

Table I.4 shows what each symbol represents.

**Table I.4**

| Symbol | Quantity |
|---|---|
| $Q$ | Charge |
| $I$ | Current |
| $V$ | Potential difference or voltage |
| $R$ | Resistance |
| $P$ | Power |
| $W$ | Work done or energy transferred |
| $\rho$ | Resistivity |
| $l$ | Length |
| $A$ | Cross-sectional area |
| $t$ | Time |

You will need to be able to recall and use the following wave formula:

$$v = f\lambda$$

where $v$ is velocity, $f$ is frequency and $\lambda$ is the wavelength.

Candidates may find the following formulae useful when answering questions in AS and A2 assessment units.

### Foundation physics mechanics formulae

$$\text{moment of force} = Fd$$
$$v = u + at$$
$$s = ut + \tfrac{1}{2}at^2$$
$$v^2 = u^2 + 2as$$
$$s = \tfrac{1}{2}(u+v)t$$
$$\text{for a spring } F = k\Delta l$$
$$\text{energy stored in a spring} = \tfrac{1}{2}F\Delta l = \tfrac{1}{2}k(\Delta l)^2$$
$$T = \frac{1}{f}$$

### Foundation physics electricity formulae

$$I = nAvq$$
$$\text{terminal p.d.} = E - Ir$$
$$\text{series circuit } R = R_1 + R_2 + R_3 + \dots$$
$$\text{parallel circuit } \frac{1}{R} = \frac{1}{R_1} + \frac{1}{R_2} + \frac{1}{R_3} + \dots$$
$$\text{output voltage across } R_1 = \left(\frac{R_1}{R_1 + R_2}\right) \times \text{input voltage}$$

### Waves and nuclear physics formulae

$$\text{fringe spacing} = \frac{\lambda D}{d}$$
$$\text{single slit diffraction minimum } \sin\theta = \frac{\lambda}{b}$$
$$\text{diffraction grating } n\lambda = d\sin\theta$$
$$\text{Doppler shift } \frac{\Delta f}{f} = \frac{v}{c}$$
$$\text{for } v \ll c$$
$$\text{Hubble law } v = Hd$$
$$\text{radioactive decay } A = \lambda N$$

### Properties of quarks

| Type of quark | Charge | Baryon number |
|---|---|---|
| up u | $+\tfrac{2}{3}e$ | $+\tfrac{1}{3}$ |
| down d | $-\tfrac{1}{3}e$ | $+\tfrac{1}{3}$ |
| $\bar{u}$ | $-\tfrac{2}{3}e$ | $-\tfrac{1}{3}$ |
| $\bar{d}$ | $+\tfrac{1}{3}e$ | $-\tfrac{1}{3}$ |

### Lepton numbers

| Particle | Lepton number L | | |
|---|---|---|---|
| | $L_e$ | $L_\mu$ | $L_\tau$ |
| $e^-$ | 1 | | |
| $e^+$ | $-1$ | | |
| $\nu_e$ | 1 | | |
| $\bar{\nu}_e$ | $-1$ | | |
| $\mu^-$ | | 1 | |
| $\mu^+$ | | $-1$ | |
| $\nu_\mu$ | | 1 | |
| $\bar{\nu}_\mu$ | | $-1$ | |
| $\tau^-$ | | | 1 |
| $\tau^+$ | | | $-1$ |
| $\nu_\tau$ | | | 1 |
| $\bar{\nu}_\tau$ | | | $-1$ |

### Geometrical and trigonometrical relationships

$$\text{circumference of circle} = 2\pi r$$
$$\text{area of a circle} = \pi r^2$$
$$\text{surface area of sphere} = 4\pi r^2$$
$$\text{volume of sphere} = \tfrac{4}{3}\pi r^3$$

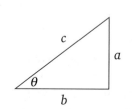

$$\sin\theta = \frac{a}{c}$$
$$\cos\theta = \frac{b}{c}$$
$$\tan\theta = \frac{a}{b}$$
$$c^2 = a^2 + b^2$$

# Unit Three Experimental Skills

## Planning

When planning an experiment you make decisions about what is worth investigating, suggest likely outcomes and devise a suitable method.

The following are some important things to remember when planning an experiment:
- identify all the variables in the experiment
- make a prediction by devising a sensible hypothesis (theory) of how changing one quantity might affect another
- predict the graph shape you would expect if your theory were correct
- decide what variable to control and how this is to be achieved
- choose suitable magnitudes for controlled variables
- select suitable apparatus and give reasons for your choice
- decide on a suitable procedure, including the range and distribution of the readings you will make
- state any risks to persons or apparatus that are present in the experiment and explain how you would minimize the risk
- explain procedures and precautions you would make to ensure your data are as accurate as possible
- describe the tabulation of the data
- explain how you would use the data to test your theory.

## Implementing

When you carry out an experiment you are being tested on your ability to use skills common to all practical work and the particular skills related to the type of practical you are doing.

Particular skills will be learned in your day-to-day practical work during the course. Remember the following skills, which are common to all practical work:
- ensure that you work safely
- arrange the apparatus sensibly so that it is easy to make the required measurements
- consider how you can improve the accuracy of the readings you make and be prepared to state and explain how you did this (e.g. minimize parallax errors)
- make measurements in a suitable range and distribute your readings evenly in the range
- repeat measurements and calculate the average (this should normally be done after resetting the apparatus in some way)
- record all your measurements in a table, including repeated readings even if they turn out to be the same
- remember to include the quantity and its unit at the head of every column in the table and include in the table any quantities you derive from your raw data
- quote all your measurements to a consistent accuracy, remembering that the last figure reflects the accuracy of your reading, i.e. 0 is as important as 1, 2, 3, etc., for example 0.233 m and 0.200 m are consistent, 0.233 m and 0.2 m or 0.20 m are not.

## Analysing

In the long question you will obtain data and be required to analyse this graphically. You will also need to be able to analyse data without drawing a graph.

### Analysing using graphs

When analysing data graphically think what plot would produce a linear graph. It is easier to draw best lines through linear plots than through plots that suggest curves.

For example, a graph of period $T$ against length $l$ for a pendulum produces a curve and shows that as $l$ increases so does $T$. However, the graph is not a straight line. This would be true only if $T \propto l$.

Since $T = 2\pi\sqrt{\dfrac{l}{g}}$ the graph to plot is $T$ ($y$-axis) against $\sqrt{l}$ ($x$-axis) or, better still, $T^2$ ($y$-axis) against $l$, since $l$ is the actual quantity that has changed (the independent variable). Both these graphs should be straight lines through the origin.

### Drawing good graphs

- Use a sharp pencil.
- Label both axes with the quantity, the power of 10 and, where appropriate, the unit.
- Choose a sensible scale.
- Indicate the scale reading every 1 cm or every 2 cm.
- Plot points accurately and clearly.
- Draw the best line (curve or straight line).

### Scales

Use 1 cm to represent 1 unit, 2 units or 10 units. Using 1 cm to represent 3 units makes plotting 0.1 very difficult.

Mark the scale often enough along an axis to make reading the scale easy but without cluttering the axes with numbers.

### Quantity and unit

The preferred way is to write, for example, 'current/$10^{-3}$ A', meaning that each integer number on the axis represents $10^{-3}$ A.

### Make full use of graph paper

Consider using a **false origin** so that points are well spaced. A false origin is one that starts with a value other than 0. When both scales start at 0, there may be a lot of wasted graph paper that carries no information.

### Plotting points and the best line

+ or × symbols are best for plotting points but ⊙ may be used.

The best line is the one that gives as even a distribution of points about the line as possible.

When data lead to a curve it is difficult to draw the best curve through the points. Instruments such as flexi-curves, universal curves and flexible rules are useful tools.

### Measuring the gradient of a straight-line graph (Figure I.1)

The gradient of a graph is the rate of change of the $y$ quantity with the $x$ quantity. Although the term 'rate' may seem to imply that the $y$ quantity changes with *time* this is not always the case.

The gradient of a straight-line graph is constant; the rate of change remains the same.

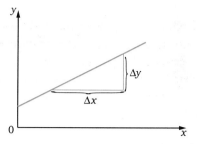

**Figure I.1**

When determining the gradient:
- use points on the line that are separated by a large distance in order to minimize errors; the points should be separated by at least half the line that has been drawn
- preferably show a large triangle on the graph or make clear in the analysis the co-ordinates of the points that are used
- always use points that are on the line. Do not use tabulated values
- determine the values of $\Delta y$ and $\Delta x$ represented by the sides of the triangle – remember to use the scales on the axes, *not* the length of the lines
- calculate the gradient, change in $y$/change in $x$, $\Delta y/\Delta x$
- remember that the value cannot be quoted to more significant figures than the readings taken from the graph for $\Delta y$ and $\Delta x$.

### Measuring the gradient of a curved graph

It is often necessary to determine the gradient at a point on the curve, as shown in Figure I.2. In this case the rate of change is varying. The gradient has to be measured at a point and gives the **instantaneous rate of change** of the $y$ quantity with the $x$ quantity.

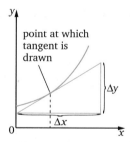

**Figure I.2**

When determining the gradient of a curved graph:
- identify the point at which the rate of change is required (e.g. when the $x$-axis is time this would be a particular time)
- draw a **tangent** at that point, i.e. a line that just touches the curve
- make the tangent line as long as is convenient to minimize errors
- using the tangent line proceed as from bullet point 4 for measuring the gradient of a straight line.

## Testing a graph for an exponential

The only exponential change you need to know for AS is the one that occurs in radioactive decay. In the practical examination you may have to test other data to check whether or not they represent an exponential change. An example of an exponential graph is shown in Figure I.3.

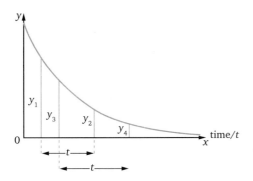

**Figure I.3**

To check whether the curve in Figure I.3 is exponential:
- take any point on the graph and determine the value of the y quantity ($y_1$)
- decide on a suitable time interval – the actual time interval does not matter ($t$)
- determine the value of the y quantity after this interval ($y_2$)
- calculate $y_2/y_1$
- repeat the procedure starting with a different value for the y quantity ($y_3$) and read the value of the y quantity ($y_4$) using the same time interval $t$
- compare the ratio $y_4/y_3$ with $y_2/y_1$ (for an exponential change the value should be the same)
- repeat the process at least once to check.

(In Figure I.3 the x quantity is time but it could be some other quantity, e.g. distance or temperature.)

Remember that in all tests using practical data there will be experimental errors. Do not therefore expect exactly the same number each time even when the theory is true. You need to make reference to any discrepancies when drawing conclusions.

## Analysing without graphs

It is possible to check whether data agree with a particular law without drawing a graph.

You should check at least three sets of data to have enough evidence to suggest that within experimental limits the law is obeyed.

For example:
- You could check whether the relationship between voltage and current is $V \propto I$. In this case $V/I$ is constant
- You could check whether frequency is proportional to the square root of the tension for vibrations of a stretched string, $f \propto \sqrt{T}$. In this case $f/\sqrt{T}$ is constant.

The relationship suggested might be one of **inverse proportionality**.

For example:
- You could check whether the resistance of a wire is inversely proportional to the square of the diameter of the wire, $R \propto 1/d^2$. In this case $Rd^2$ is constant.
- You could check whether the fringe spacing is inversely proportional to the slit separation, $y \propto 1/d$. In this case $yd$ is constant.

EXAMPLE ONE: The data in Table I.5 were collected for the tension $T$ and the corresponding frequency $f$ of vibration for a guitar string. Test whether the data agrees with the theory that $f \propto \sqrt{T}$.

**Table I.5**

| Tension/N | 150 | 115 | 175 |
|---|---|---|---|
| Frequency/Hz | 250 | 220 | 270 |

The theory suggests that $f = $ constant $\times \sqrt{T}$, therefore

$$\frac{f}{\sqrt{T}} = \text{constant}$$

Dividing each frequency by the square root of the tension gives values of 20.4, 20.5 and 20.4. These values are the same, within the limits of experimental errors.

## Checking data for an exponential change

You need three or four sets of data for which the $x$ increments are equal.

Divide the second $y$ reading by the first, the third by the second, and so on.

If the ratios are equal then the data suggest an exponential change.

EXAMPLE TWO: The data in Table I.6 show the amplitude of an oscillation at 20 s intervals. Check whether or not the amplitude decays exponentially with time.

**Table I.6**

| Time/s | 0 | 20 | 40 |
|---|---|---|---|
| Amplitude/mm | 58 | 41 | 34 |

$\frac{41}{58} = 0.71$, $\frac{34}{41} = 0.83$

These values are very different even allowing for experimental uncertainties so on this evidence the change is *not* exponential.

## Evaluating

Evaluating means answering the question 'How good was the experiment?'. This requires you to:
- identify the strengths and weaknesses of the experiment
- use your knowledge of alternative procedures and apparatus to describe and explain improvements to the experiment that you have been asked to evaluate
- determine **uncertainties** in the raw data and in the final result.

The first two points will be learned in your day-to-day practical work on particular topics. For example you might identify:
- problems with the precision of apparatus
- difficulties in making a particular measurement because of apparatus arrangements.

You might suggest:
- the use of different instruments for a length measurement, such as a micrometer or vernier callipers
- data capture equipment that would produce a more accurate result because of ease of measurement or a better procedure.

## Uncertainties

An important part of any experiment is assessing any **uncertainty** in measurements and in the final result. The uncertainty in a measurement defines the range within which you suggest the true value lies.

Measurements made in scientific experiments are never absolutely precise. Any measured value has an uncertainty. The lower the uncertainty, the more accurate the measurement.

Possible sources of uncertainty are:
- the design of the instrument
- the calibration of the instrument
- the ability of the user to read the instrument accurately
- the design of the experiment (setting up and procedure) in which the instrument is used.

In a good experiment care must be taken to reduce each source of error to a minimum.

### Estimating the uncertainty in a reading

When using instruments the smallest scale division is a good guide for estimating the uncertainty. Assuming correct calibration a reading can be made to $\pm 1$ scale division, e.g. using a metre ruler, a length can be measured to an accuracy of $\pm 1$ mm or using a micrometer a length can be measured to $\pm 0.01$ mm.

### Absolute uncertainty

A length measured as $124 \pm 1$ mm means that the length lies somewhere in the range 123 mm to 125 mm. 1 mm in this case is the absolute uncertainty in the length.

### Percentage uncertainty

Uncertainties may be quoted as percentages rather than absolute values. An uncertainty of $124 \pm 1$ means 1 in 124, i.e.

percentage uncertainty $= \frac{1}{124} \times 100 \approx 0.8\%$

### Use of significant figures

When a length is written as 124 mm there is an implied uncertainty in the last figure which is assumed to be $\pm 1$ mm. If you want to express a greater uncertainty then you need to state it.

### Combining uncertainties

To derive a quantity from a set of measurements, quantities are substituted for symbols in a formula. The value you obtain using the formula has an uncertainty. This depends on the uncertainties of each measurement substituted.

Care has to be taken when combining uncertainties. Sometimes absolute values are used and sometimes percentage uncertainties are used.

#### Adding and subtracting

Total uncertainty = sum of absolute uncertainties.

#### Multiplying and dividing

Total uncertainty = sum of percentage uncertainties.

#### Squaring and cubing

When a measured quantity is squared, the quantity is multiplied by itself, so the percentage uncertainty is *twice* the uncertainty in the measured quantity.

The percentage uncertainty in a cubed quantity is *three times* the uncertainty in the measured quantity.

In practical work it is important to take these factors into account when deciding how to measure quantities that will be squared or cubed. For example, you will need to use instruments or methods that provide greater precision when measuring small quantities, such as the diameter of wire.

### Effect of repeat measurements

When you repeat readings you reduce the uncertainty. The uncertainty should be calculated from statistics but the following is good enough for AS (and A2) purposes.

#### Uncertainty for n measurements of the same quantity

The absolute uncertainty in the mean (or average) value is given by:

$$\text{uncertainty} = \frac{\text{maximum value} - \text{minimum value}}{n}$$

It follows from this that the more readings you take the lower the uncertainty will be, i.e. you will improve the accuracy of the measurement.

**EXAMPLE THREE:** The height of student A is $1.45\pm0.01$ m and the height of student B is $1.36\pm0.01$ m. The difference in their heights is:

$$(1.45\pm0.01) - (1.36\pm0.01) = 0.09\pm0.02$$

Notice that the percentage uncertainty in the difference ($\approx 20\%$) is much greater than the percentage uncertainty in each value ($\approx 0.7\%$).

Looked at another way, if the true height of student A was at the high end of the range and the height of student B was at the low end, the difference could have been $1.46 - 1.35 = 0.11$ m. The smallest possible difference is $1.44 - 1.37 = 0.07$ m.

**EXAMPLE FOUR:** In an experiment to measure resistance, the following data together with their uncertainties were obtained:

potential difference = $V = (3.5\pm0.1)$ V
current = $I = 1.48\pm0.02$ A
resistance = $V/I = 2.36$ Ω

% uncertainty in $V \approx (0.1/3.5) \times 100 = 2.9\%$
% uncertainty in $I \approx (0.02/1.48) \times 100 = 1.35\%$
% uncertainty in $R \approx 2.9 + 1.35 = 4.25\%$

The absolute uncertainty in $R$ is:

$\frac{4.25}{100} \times 2.36 \approx 0.10$ Ω

Resistance = $(2.36\pm0.10)$ Ω

Since uncertainties are themselves estimates and the potential difference measurement is given to two significant figures, the final answer should be quoted as $(2.4\pm0.1)$ Ω.

Taking a second measurement using a different voltage and averaging the two would provide a more accurate value.

## Questions on practical skills

For each question consider the points above and devise a detailed plan to investigate any one of them.

### Planning

**1** A cylinder is rolled down an inclined plane as shown in Figure I.4. The time taken to reach the bottom is to be measured.

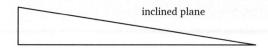

**Figure I.4**

For this exercise write down all the variables, including the properties of the cylinder itself, that might determine the time measured.

**2** The apparatus shown in Figure I.5 is used to determine the resistance of a liquid (copper sulphate solution).

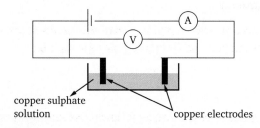

**Figure I.5**

For this exercise you must consider all the variables that might affect the resistance between the electrodes. You will also need to decide what meters to use. How will you decide?

### Implementing

You could try out your plan for question 1 with equipment you have at home. Otherwise your teacher may let you try it out with better equipment in the laboratory.

You would need to work in a laboratory to carry out your plan for question 2.

### Analysing and evaluating

**3** The data in Table I.7 were recorded by a student in an oscillator experiment. Masses were added to a spring and the period measured for each mass. The student suggested the hypothesis that the period is directly proportional to the square root of the total extension that each mass produces when it has come to rest.

**Table I.7**

| Extension, $l$ (m) | 0.05 | 0.11 | 0.15 | 0.21 | 0.25 |
|---|---|---|---|---|---|
| Time taken for five oscillations/s | 2.4 | 3.1 | 3.8 | 4.6 | 5.3 |

(a) Plot a graph of period against the square root of the extension.
(b) Explain whether or not the data supports the hypothesis.
(c) The equation for the period $T$ is

$$T^2 = 4\pi^2 \frac{l}{g}$$

(d) Use the set of data for an extension of 0.25 m to obtain a value for $g$.
(e) Determine the uncertainty in this value assuming that the data obtained are all $\pm 1$ in the final figure.
(f) Explain what precautions you would take and how you would modify the procedure to obtain more accurate data.

**4** In an experiment to measure an unknown resistance a student uses the experimental arrangement shown in Figure I.6.

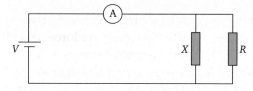

**Figure I.6**

The student obtained the data in Table I.8 for the variation of current for different values of the resistance $R$. The current was measured with a meter with a full-scale deflection of 200 mA. Nominal values of $R$ were determined from the colour coding. The tolerance for these nominal values was 10%.

**Table I.8**

| Resistance $R/\Omega$ | 100 | 220 | 330 | 470 |
|---|---|---|---|---|
| Current $I$/mA | 36 | 28 | 14 | 11 |

The student deduces that the equation for the circuit is:

$$I = \frac{V}{R} + \frac{V}{X}$$

(a) Plot a graph of $I$ against $1/R$.
(b) (i) Determine the gradient of the graph.
    (ii) Determine a value for $V$.
    (iii) Determine the value of $X$.
(c) State the safety precautions you would take if you were carrying out the experiment.
(d) State and explain what changes you would make to the apparatus or the procedures to determine a more accurate value for $X$.

# Unit 1: Scalars and Vectors

## Scalar Quantities

Direction has no meaning for scalar quantities. However, they do have magnitude or size. Examples of scalar quantities are **mass**, **energy** and **charge**. When you add two scalar quantities together, you do not have to think about direction. Two 4.0 kg masses always have a combined mass of 8.0 kg.

## Vector Quantities

For vector quantities, direction is important. Examples of vector quantities are **force**, **velocity** and **acceleration**. You can always specify a direction for these quantities, for example a 5.0 N force to the right or a velocity of 25 m s$^{-1}$ to the north. **Weight** is another vector quantity because it is an example of a force. The direction of weight is always downwards.

When adding vectors, you need to take direction into account. In Figure 1.1 two forces are added together.

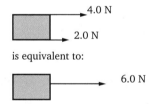

**Figure 1.1**

Notice that the length of the arrow is an indication of the magnitude of the force.

In Figure 1.2, the two forces are acting in opposite directions so the overall, or resultant, force is 2.0 N to the right. You should get used to stating the direction when you describe a force.

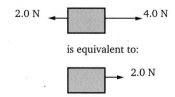

**Figure 1.2**

## Adding Forces at Right Angles

Clearly, the object in Figure 1.3 will be pulled upwards and to the right.

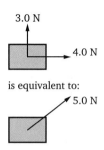

**Figure 1.3**

To find the size and direction of the resultant force you can use a scale drawing or a calculation.

## Scale Drawing

To make a scale drawing follow these instructions (the steps are shown in Figure 1.4).
- Draw the forces to scale, for example 5 mm for each newton.
- Slide one of the forces along the other force, keeping its direction the same (in this case up) so that the arrows on the two forces follow on from each other.
- Complete the triangle of forces from the start of one force to the end of the other.
- Find the magnitude of the resultant force (*F*) by measuring the length of the diagonal arrow and using your scale (5.0 N in this case).
- Find the direction of the resultant force by measuring the angle ($\theta$) with a protractor (37° from the horizontal in this case).

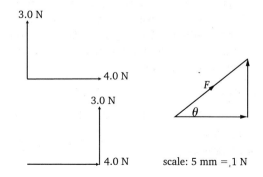

**Figure 1.4**

## Calculation

- Draw the triangle of forces in the same way as above, but not necessarily to scale.
- Use Pythagoras's theorem to find the magnitude of the resultant force, which is represented by the hypotenuse of the triangle of forces.
- Use algebra to find the angle at which the resultant force acts.

In this case, from Figure 1.4:

$$F^2 = 3^2 + 4^2$$
$$= 9 + 16$$
$$= 25$$
$$F = 5.0 \text{ N}$$
$$\tan \theta = \tfrac{3}{4}$$
$$\theta = 37°$$

**Test your understanding**

1. Use scale drawings to find the resultant forces on each body. Remember to quote the scale that you use and include the direction of the resultant force in your answer.

    (a)
    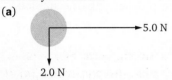
    **Figure 1.5**

    (b)
    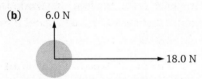
    **Figure 1.6**

    (c)
    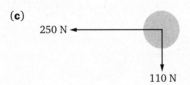
    **Figure 1.7**

    (d)
    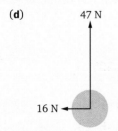
    **Figure 1.8**

2. Repeat question 1 using calculation instead of scale drawings.

## Addition of Other Vector Quantities

All vector quantities can be added. In particular, it is often necessary to add components of velocity to find a resultant.

## Resolving Vector Quantities

It is sometimes necessary to split forces up into two **components** at right angles to each other. For example, in Figure 1.9 a car is being towed. In order to find out how much of the force in the tow rope is being used to pull the car along the road and how much is being used to lift the front of the car, the towing force should be **resolved** into its horizontal and vertical components.

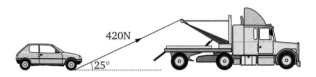

**Figure 1.9**

To do this by drawing, complete a rectangle or triangle of forces to scale and measure the two components (Figure 1.10).

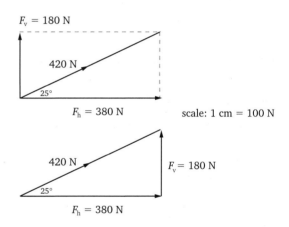

**Figure 1.10**

To achieve the same by calculation, sketch the triangle of forces and use algebra to find the magnitudes of the two components (Figure 1.11).

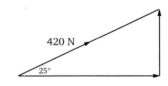

**Figure 1.11**

$$\frac{F_v}{420} = \sin 25° \qquad \frac{F_h}{420} = \cos 25°$$
$$F_v = 420 \sin 25° \qquad F_h = 420 \cos 25°$$
$$F_v = 180 \text{ N} \qquad F_h = 380 \text{ N}$$

### Test your understanding

1. An aircraft is directed due north at a speed of 240 m s$^{-1}$. A side wind is blowing from the west at 35 m s$^{-1}$. Find the resultant speed and course of the aircraft.
2. A ship is steaming south at a speed of 12 knots in a tidal stream that is moving to the west at 3.5 knots. Find the resultant speed and direction of the ship.
3. Find the resultant horizontal component of the force and then find the overall resultant force.

   (a)
   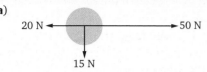

   **Figure 1.12**

   (b)
   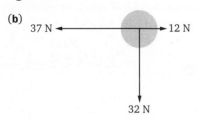

   **Figure 1.13**

   (c)
   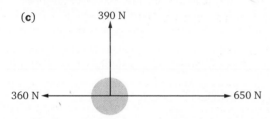

   **Figure 1.14**

4. Find the horizontal and vertical components of these forces:

   (a)
   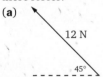

   **Figure 1.15**

   (b)
   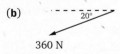

   **Figure 1.16**

   (c)
   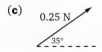

   **Figure 1.17**

## Resolution of Velocities

Velocities may be resolved in the same way as forces. This is often useful in order to calculate the time taken for an event. For example, how long does it take an aircraft, climbing at an angle of 15° and flying at a speed of 250 m s$^{-1}$, to cover a horizontal distance of 16 km?

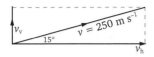

**Figure 1.18**

$$v_h = 250 \cos 15°$$
$$= 241 \text{ m s}^{-1}$$

$$\text{time} = \frac{\text{distance}}{\text{speed}}$$
$$= \frac{16\,000}{241}$$
$$= 66 \text{ s}$$

## Resolving Forces in Other Directions

When dealing with sloping surfaces, it is often useful to resolve forces into components parallel with and at right angles to the surface. The principle is the same as for resolving horizontally and vertically; it is just necessary to be careful in working out the angles.

Figure 1.19 shows a body of weight 4.5 N on an inclined plane. To work out the motion of the body, it is necessary to find the component of the weight, $W$ of the body parallel to the plane.

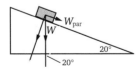

**Figure 1.19**

$$\frac{W_{par}}{W} = \sin 20°$$

$$W_{par} = W \sin 20°$$
$$= 4.5 \sin 20°$$
$$= 1.5 \text{ N}$$

### Test your understanding

1. A projectile is fired at an angle of 25° upwards from the horizontal at a speed of 30 m s⁻¹. Calculate:
   (a) its horizontal component of velocity
   (b) its vertical component of velocity.

2. A river is 240 m wide. Two canoes are crossing the river at an angle to the banks. Canoe A is paddled at a speed of 5.0 m s⁻¹ at an angle of 50° to the bank and canoe B is paddled at a speed of 4.0 m s⁻¹ at an angle of 30° to the bank. The canoes set off together.
   (a) Calculate the components of velocity at right angles to the bank for each canoe.
   (b) Which canoe gets to the opposite bank first, and by how much?
   (c) Calculate the components of velocity parallel to the bank for each canoe.
   (d) Determine the distance between the landing points of the two canoes.

3. For the following bodies, find the components of the weights parallel to and perpendicular to the planes.

   (a)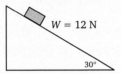

   **Figure 1.20**

   (b)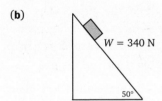

   **Figure 1.21**

## Calculating Moments

When more than one force acts on a body, it can have the effect of turning the body. The size of the turning effect that a force has obviously depends on the magnitude of the force. The turning effect or **moment** of the force also depends on the distance between the force and the turning point (or **pivot**).

$$\text{moment of a force} = \text{force} \times \text{distance between force and pivot}$$

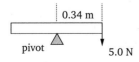

**Figure 1.22**

$$\text{moment} = 5.0 \times 0.34 = 1.7\,\text{Nm}$$

Notice that the unit of moment is N m.

When the force acts at a different angle, the distance that should be used in the calculation of the moment is the **perpendicular distance** between the force and the pivot (Figure 1.23).

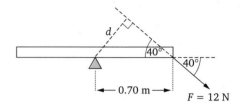

**Figure 1.23**

In this case, the perpendicular distance between the force and the pivot is $d$ where:

$$\frac{d}{0.70} = \sin 40°$$

$$\begin{aligned}\text{moment} &= F \times d \\ &= F \times 0.70 \times \sin 40° \\ &= 12 \times 0.70 \times \sin 40° \\ &= 5.4\,\text{Nm}\end{aligned}$$

You will not be required to calculate perpendicular distances in the above example. The perpendicular distance will be given when necessary.

### Test your understanding

4. Calculate the moments of the forces shown in the following diagrams.

   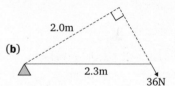

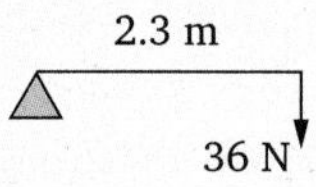

   **Figure 1.24**

   **Figure 1.25**

When a body is in equilibrium, the net moment acting on it is zero:

$$\text{the sum of the clockwise moments} = \text{the sum of the anticlockwise moments}$$

This equation can be used to find the missing force in Figure 1.26.

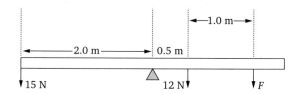

**Figure 1.26**

Taking moments about the pivot P:

$$\Sigma \text{ (anticlockwise moments)} = \Sigma \text{ (clockwise moments)}$$

$$(15 \times 2.0) = (12 \times 0.5) + (F \times 1.5)$$

$$30 = 6.0 + 1.5F$$

$$1.5F = 24$$

$$F = 16 \text{ N}$$

It is not always necessary to take moments about a pivot. In equilibrium, the net moment about *any* point in the body is zero. It is often convenient to choose a different point about which to take moments.

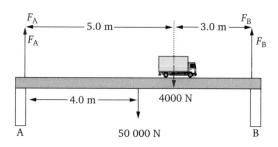

**Figure 1.27**

In Figure 1.27, it is required to find the force ($F_B$) in pillar B, which holds up the right-hand end of the bridge. By choosing to take moments about A, the force $F_A$ becomes irrelevant to the calculation. Since the moment of force is found by multiplying the force by the distance to the point, there is no moment of force $F_A$ about A.

Taking moments about A:

$$\Sigma \text{ (clockwise moments)} = \Sigma \text{ (anticlockwise moments)}$$

$$(4000 \times 5.0) + (50\,000 \times 4.0) = (F_B \times 8.0)$$

$$20\,000 + 200\,000 = 8.0 F_B$$

$$F_B = \frac{220\,000}{8.0}$$

$$F_B = 27\,500 \text{ N}$$

**Test your understanding**

1. In Figure 1.28 find the distance between the counterweight (L) and the pivot (P) so that the crane jib is in equilibrium. The weight of the jib itself acts at the point G, 3.0 m from the pivot.

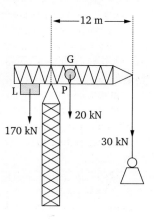

**Figure 1.28**

2. Figure 1.29 shows a lighting bar of mass 15 kg. It is suspended from the ceiling by two metal bars, A and B. Two lamps are suspended from the bar. The masses and positions of the lights are shown in the figure ($g = 9.8 \text{ m s}^{-2}$).

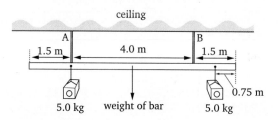

**Figure 1.29**

(a) State the principle of moments.
(b) Calculate the tensions in the bars A and B.

3. Figure 1.30 shows the apparatus used for finding the viscosity of a liquid. A ball-bearing is dropped into the liquid. The time it takes to move between two rubber bands separated by a known distance is measured.

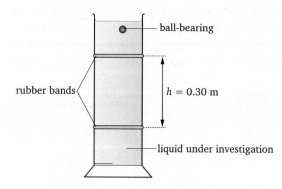

**Figure 1.30**

continued ▷

Figure 1.31 shows how the velocity of the ball-bearing changes over the first 100 ms after it is released.

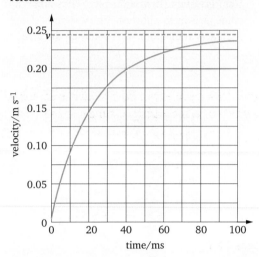

**Figure 1.31**

(a) Use the graph to find the initial acceleration of the ball-bearing.
(b) Explain why the acceleration of the ball-bearing is not constant.
(c) Explain why the velocity of the ball-bearing becomes constant at the value marked v on the graph.
(d) Use the graph to find the distance travelled by the ball-bearing in the first 50 ms.

4  A train has a mass of 80 000 kg. Using 75% of its maximum available force, it accelerates from rest to a speed of 18 m s$^{-1}$ in 24 s and continues at this speed for 100 s. It then slows steadily to a stop in 30 s.
   (a) Calculate:
       (i)   the acceleration and deceleration of the train
       (ii)  the distance travelled during the journey
       (iii) the force used to accelerate the train
       (iv)  the force that would be required to slow the train during its deceleration.
   (b) When it sets off again, the train is going up an incline at angle of 3.0°. Calculate:
       (i)  the magnitude of the component of its weight down the slope
       (ii) the maximum acceleration it can achieve up the slope.

# Kinematics

## Velocity and Speed

A body's velocity describes its speed and its direction. Speed is simply the magnitude of the velocity vector. Speed is defined as distance travelled per unit time. The symbol $v$ is used to denote speed.

$$v = \frac{s}{\Delta t}$$

where $s$ = distance and $\Delta t$ = time taken.

The SI unit of speed is m s$^{-1}$.

## Distance–time Graphs

The graphs in Figure 1.32 denote bodies moving with constant speed.

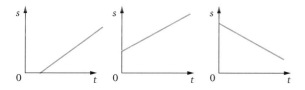

**Figure 1.32**

Since all of these graphs are straight lines (with constant gradients), they all show bodies with constant speed.

To find the speed of a body from its distance-time graph, simply find the gradient of the line, e.g. for the graph in Figure 1.33:

$$\text{gradient} = v = \frac{s}{\Delta t}$$
$$= \frac{28}{7.0}$$
$$= 4.0 \text{ m s}^{-1}$$

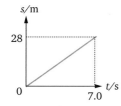

**Figure 1.33**

The graphs in Figures 1.34 and 1.35 denote bodies with changing speed.

Figure 1.34 represents a body which is accelerating: the gradient is increasing as time goes on.

**Figure 1.34**

Figure 1.35 represents a body which is decelerating: the gradient of the graph is decreasing as time goes on.

**Figure 1.35**

To find the speed from a distance–time graph that is a curve, draw a tangent to the curve at the desired point and find the gradient of the tangent (Figure 1.36).

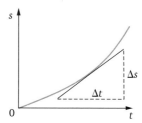

**Figure 1.36**

$$\text{speed} = \frac{\Delta s}{\Delta t}$$

## Acceleration

Acceleration is defined as the change in velocity per unit time.

$$a = \frac{\Delta v}{\Delta t}$$

where $\Delta v$ = change in velocity and $\Delta t$ = time taken.

The SI unit of acceleration is m s$^{-2}$.

## Speed–time Graphs

The graphs in Figure 1.37 represent bodies with constant acceleration.

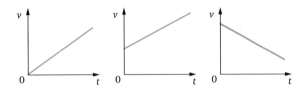

**Figure 1.37**

Since these are straight-line graphs (with constant gradients), they all show constant acceleration. Notice that the third of these graphs has a negative gradient. This indicates that the body represented by the graph is slowing down. Deceleration is an alternative term for negative acceleration.

Acceleration is found from the speed–time graph by determining the gradient.

## Finding Distance Travelled from a Speed–time Graph

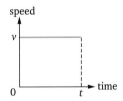

**Figure 1.38**

Figure 1.38 represents a body travelling at a constant speed, $v$, for a time $t$:

$$\text{distance} = vt$$

Note that the distance travelled is equivalent to the area under the graph. The area of a rectangle is found by length × breadth.

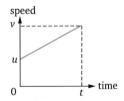

**Figure 1.39**

Figure 1.39 represents a body accelerating from a speed of $u$ to a speed of $v$ in a time $t$. The body has an average speed of $\frac{1}{2}(u + v)$:

$$\text{distance} = \text{average speed} \times \text{time}$$
$$= \tfrac{1}{2}(u + v)t$$

Notice that this is also equivalent to the area under the graph. The area of a trapezium is the average length of the parallel sides times their separation.

For all speed–time graphs:

$$\text{distance travelled} = \text{area under the graph}$$

### Test your understanding

1 Find the speeds of the bodies represented by the following graphs.

(a)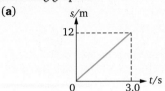

**Figure 1.40**

(b)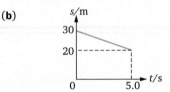

**Figure 1.41**

2 Find the accelerations of the bodies represented by the following graphs.

(a)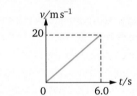

**Figure 1.42**

(b)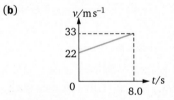

**Figure 1.43**

3 Find the distances travelled by the bodies represented by the graphs in question 2. Do the calculations by using the equations of motion *and* by finding the areas under the graphs.

4 (a) Find the distance travelled by the body represented in Figure 1.44. You cannot use the equations of motion as the acceleration is not constant.

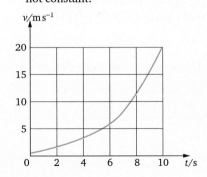

**Figure 1.44**

(b) Find the acceleration of the body at time $t = 6$ s.

## Finding the Acceleration Due to Gravity, g

A data logger may be used to determine the acceleration of a body falling freely under gravity (Figure 1.45).

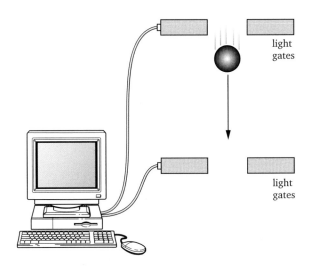

**Figure 1.45**

The size of the body being dropped is entered into the computer. The computer measures the time taken for the body to pass through the top set of light gates. It calculates the body's speed at the top set of gates and again at the lower gates. It also measures the time interval between the gates.

The computer then determines the acceleration using:

$$a = \frac{\text{change in speed}}{\text{time taken}}$$

The acceleration due to gravity can also be found by timing the oscillations of pendulums or of mass/spring systems. Details of these can be found on pages 39 and 40.

## Equations of Motion

The following equations are useful in the analysis of the movement of bodies having uniform motion. They apply only to situations where the acceleration is constant.

$$s = vt$$

where $s$ = distance, $v$ = average or constant speed and $t$ = time.

$$v = u + at$$
$$s = ut + \tfrac{1}{2}at^2$$
$$v^2 = u^2 + 2as$$
$$s = \tfrac{1}{2}(u + v)t$$

where $v$ = final speed, $u$ = initial speed and $a$ = acceleration.

It is necessary to know these equations and to be familiar with their use but you will not be expected to be able to derive them. It is necessary to practice manipulating the equations so their derivations are given below.

## Derivation of the Equations of Motion

$$v = u + at$$

$$\left[ a = \frac{\Delta v}{t} \right]$$

but $\Delta v = v - u$

so $a = \dfrac{v - u}{t}$

Multiply both sides by $t$:

$$at = v - u$$

Rearranging gives:

$$v = u + at$$

$$s = \tfrac{1}{2}(u + v)t$$

See page 24.

$$s = ut + \tfrac{1}{2}at^2$$

$$s = \tfrac{1}{2}(u + v)t$$

but $v = u + at$

so $s = \tfrac{1}{2}(u + u + at)t$

$s = \tfrac{1}{2}(2u + at)t$

$s = ut + \tfrac{1}{2}at^2$

$$v^2 = u^2 + 2as$$

$$s = \tfrac{1}{2}(u + v)t$$

but $v = u + at$

or $t = \dfrac{(v - u)}{a}$

so $s = \tfrac{1}{2}(u + v) \times \dfrac{(v - u)}{a}$

Multiply both sides by $2a$:

$$2as = (v + u)(v - u)$$
$$2as = v^2 - u^2$$

or $v^2 = u^2 + 2as$

## Using the Equations of Motion

The key strategy when using the equations of motion is to organize your data to start with. You should then be able to identify which equation to use. The examples that follow illustrate this.

**EXAMPLE ONE:** A stone, originally at rest, falls from a 75 m high cliff. How long does it take to fall? The acceleration due to gravity, g, is 9.8 m s$^{-2}$.

$$s = 75 \text{ m}$$
$$u = 0 \text{ m s}^{-1}$$
$$a = g = 9.8 \text{ m s}^{-2}$$
$$t = ?$$

Inspection of the data shows that the equation that involves the known parameters and the one we want to find is:

$$s = ut + \tfrac{1}{2}at^2$$

Substituting the data gives:

$$75 = (0 \times t) + \tfrac{1}{2}(9.8)t^2$$
$$= 4.9t^2$$

since $(0 \times t) = 0$

$$t^2 = \frac{75}{4.9}$$
$$t = 3.9 \text{ s}$$

**EXAMPLE TWO:** A car accelerates from 13 m s$^{-1}$ to 30 m s$^{-1}$ in 6.0 s. Calculate the distance travelled by the car as it accelerates.

$$v = 30 \text{ m s}^{-1}$$
$$u = 13 \text{ m s}^{-1}$$
$$t = 6.0 \text{ s}$$
$$s = ?$$

By inspecting the data, it can be seen that none of the equations of motion is immediately applicable. Those involving $s$ also contain $a$, which is not known. Clearly it is necessary to calculate $a$ before attempting to find $s$.

$$a = \frac{v - u}{t}$$
$$= \frac{30 - 13}{6.0}$$
$$= 2.83 \text{ m s}^{-2}$$

It is now possible to identify an equation of motion that can be used:

$$s = ut + \tfrac{1}{2}at^2$$
$$= (13 \times 6.0) + \tfrac{1}{2}(2.83)6.0^2$$
$$= 129 \text{ m}$$

(or 130 m to two significant figures)

### Test your understanding

1. A car accelerates from rest to 20 m s$^{-1}$ in 8.0 s. Calculate the acceleration of the car and the distance that it travels.
2. A sprinter accelerates from rest to 8.9 m s$^{-1}$ in the first 1.5 s of a race. She completes the 100 m race at a constant speed of 8.9 m s$^{-1}$. Calculate:
   (a) the sprinter's acceleration in the first 1.5 s
   (b) the distance covered in the first 1.5 s
   (c) the remaining distance to be covered
   (d) the time taken for the sprinter to complete the race.
3. An object is projected vertically upwards at a speed of 22 m s$^{-1}$ ($g = 9.8$ m s$^{-2}$). Calculate:
   (a) the maximum height reached by the object (the speed at the top is zero)
   (b) the time taken between the object being projected and it landing again.

## Non-uniform Acceleration

The gradient of a distance–time graph at any point is equal to the speed at that point, even when acceleration is changing. This lets you calculate the speed, at any time, in the oscillating system shown on page 39.

The speed at time $t$ is the gradient of the tangent of the curve at that time (Figure 1.46).

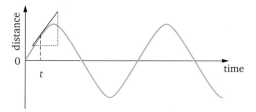

**Figure 1.46**

The gradient of the speed–time graph at any point is equal to the acceleration at that point, even when acceleration is changing. This lets you calculate the acceleration, at any time, in the oscillating system shown on page 39.

The acceleration at time $t_1$ is the gradient of the tangent of the curve at that time (Figure 1.47).

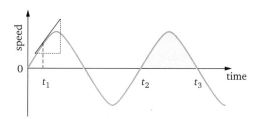

**Figure 1.47**

For varying accelerations, the distance travelled by a body is still equivalent to the area under the speed–time graph. For example, the distance travelled between times $t_2$ and $t_3$ on Figure 1.47 is equivalent to the shaded area.

As the shape of the graph is unusual, finding the area is best done by counting squares on graph paper calculating the distance represented by a square, and multiplying them together.

## Bodies Moving Horizontally and Vertically at the Same Time

A tennis ball hit horizontally when it is served will drop as it travels (Figure 1.48).

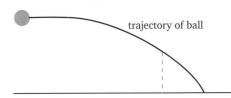

**Figure 1.48**

Situations like this can be analysed by recognizing that the horizontal and vertical components of the motion can be treated separately.

No forces act on the ball in a horizontal direction (if air resistance is ignored). The ball's horizontal component of velocity does not change.

The ball's weight acts on it vertically downwards, causing a steady increase in the vertical component of velocity.

For example, if the initial horizontal velocity of the ball is known to be 39 m s$^{-1}$, and it is hit at a height of 3.0 m, the horizontal distance travelled before it hits the ground can be found as follows.

Consider the vertical component:

$$s = ut + \tfrac{1}{2}at^2$$

vertical distance  $s = 3.0$
initial vertical speed  $u = 0$
vertical acceleration  $a = g = 9.8$ m s$^{-2}$

$$3.0 = 0 + \tfrac{1}{2}(9.8)t^2$$
$$t = 0.78 \text{ s}$$

Now consider the horizontal component:

$$s = vt$$

horizontal velocity  $v = 39$ m s$^{-1}$
time taken  $t = 0.78$ s

$$s = 39 \times 0.78$$
$$= 30 \text{ m}$$

### Test your understanding

1. A ball is thrown from ground level so that it has a vertical component of velocity of 12 m s$^{-1}$. Calculate:
   (a) the time taken for the ball to reach its highest point
   (b) the total time for which the ball is in the air
   (c) the maximum height reached by the ball.
   (d) The ball also has a horizontal component of velocity of 15 m s$^{-1}$. Calculate the horizontal distance travelled before it hits the ground ($g = 9.8$ m s$^{-2}$).

2. For the projectile shown in Figure 1.49 calculate the maximum height reached and the horizontal distance travelled before it reaches the ground.

**Figure 1.49**

3. A nut falls from the underside of a car travelling at 30 m s$^{-1}$. The nut falls from a point that is 0.35 m above the road surface. Calculate:
   (a) the horizontal distance travelled by the nut before it hits the road
   (b) the angle at which the nut hits the road.

# Springs, Energy and Work

When a spring is stretched within its elastic limit, the extension ($\Delta l$) is proportional to the applied load ($F$).

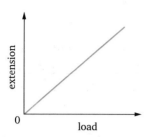

**Figure 1.50**

The relationship between the extension and the load is:

$$F = k\Delta l$$

where $k$ is a constant for any particular spring and is called the spring stiffness. The unit for $k$ is $\mathrm{N\,m^{-1}}$.

It is common, when working with springs, to measure $\Delta l$ in centimetres, so you should remember to convert this to metres before calculating $k$.

## Energy Stored in a Stretched Spring

As a spring is loaded, work is done. The energy used to do this work is stored in the spring:

$$\text{work done} = \text{force} \times \text{distance moved}$$

When the load is small, a 1 cm extension requires little work to be done.

For larger loads, the force used to do the stretching is bigger so another 1 cm extension will need a larger amount of work to be done.

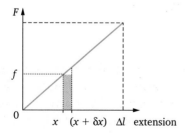

**Figure 1.51**

To extend the length of the spring by a small amount $\delta x$ from $x$ to $(x + \delta x)$, a small amount of work, $\delta W$ is needed. At this point, the force being used to do the extending is $f$:

$$\text{work done} = \text{force} \times \text{distance moved}$$
$$\delta W = f \times \delta x$$

$f \times \delta x$ is the area of the rectangle shaded on Figure 1.51 (ignoring the small triangle at the top). $\delta W$ is equivalent to the area under this part of the graph.

The work done stretching the spring by its whole extension $\Delta l$ is $W$:

$$W = \text{the area under the whole graph}$$

For a triangle, area = $\tfrac{1}{2}$ height $\times$ base, so:

$$W = \tfrac{1}{2} F \Delta l$$

Since $F = k\Delta l$, this equation can also be written as:

$$W = \tfrac{1}{2} k (\Delta l)^2$$

## Energy Released from a Stretched Spring

When the spring in Figure 1.52 is released, the energy stored will be transferred into kinetic energy in the ball-bearing.

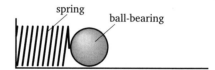

**Figure 1.52**

The principle of storing energy in springs for subsequent release is used in mechanisms from vehicle suspension systems to staple guns.

### Test your understanding

1. A spring extends by 8.0 cm when a 600 g load is added. Calculate:
   (a) the force used to extend the spring
   (b) the spring constant
   (c) the energy stored in the spring when it is fully extended.
2. (a) Find the spring constant for the spring used to produce the graph in Figure 1.53.
   (b) Calculate the force needed to fully extend the spring.

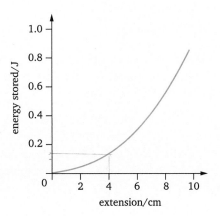

**Figure 1.53**

## Electrical Energy Sources

Fossil-fuelled power stations have traditionally been used to generate electricity. The fuels used in these power stations include coal, oil and gas.

More recently, nuclear energy has been used for electricity generation.

Other alternative power sources include the use of energy from waves, tides, wind, rivers and lakes, solar radiation, and geothermal and biomass. All of these (except solar power) involve the use of generators.

In a generator, a magnetized rotor spins near stationary coils, causing electricity to be produced in the coils.

### Electricity generation involving steam

Fossil fuel or plant material (biomass) is burned (Figure 1.54). The hot gases produce steam in boilers. High-pressure steam is used to drive turbines, which turn generators to produce electricity.

Nuclear and geothermal facilities have other ways of producing the steam but are otherwise similar.

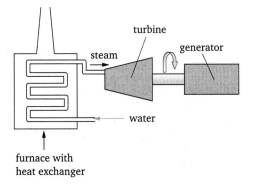

**Figure 1.54**

### Generators using moving water

Moving water is used to turn turbines in tidal (Figure 1.55) and hydro-electric (Figure 1.56) power generation.

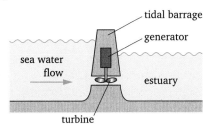

**Figure 1.55**

The turbine/generator sets produce electricity as the tide is coming in and as it is going out.

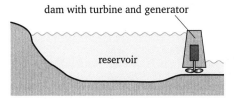

**Figure 1.56**

Hydro-electric schemes can use either small dams across rivers or larger reservoirs, which are built for the purpose or adapted from existing lakes.

### Wind power

Wind turbines are turned by the action of the wind on a propeller (Figure 1.57). The generator set turns on the pylon so that the propeller faces the direction of the wind.

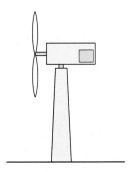

**Figure 1.57**

### Wave power

One type of device for the generation of electricity from waves involves a bobbing 'duck', which oscillates as a wave moves past. This operates a pump, which drives a turbine/generator set (Figure 1.58).

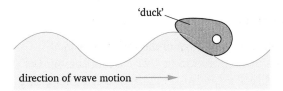

**Figure 1.58**

Another design causes the wave to compress a trapped column of air, which is then expelled from the column, causing a fan to turn as it does so. The fan operates a generator. As the wave recedes, air is drawn into the column, once again rotating the fan (Figure 1.59).

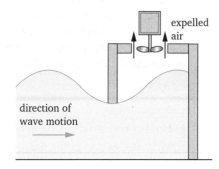

**Figure 1.59**

### Solar power

Domestic solar panels usually use infrared radiation in sunlight to heat water. They are not used for electricity generation.

Electricity can be produced from sunlight using photo-voltaic panels. In principle, these are the same as the solar cells used to power calculators. Large arrays of such cells can produce significant quantities of electricity, especially if they are mounted so that they can turn to follow the sun.

## Choosing How to Produce Electricity

The following need to be considered when choosing a method of electricity generation:
- capital cost
- fuel costs
- operating and maintenance costs
- the availability of skilled labour and spare parts
- decommissioning costs
- environmental issues, including air, ground and water pollution, damage to habitats, visual and noise pollution, the need for infrastructure to service installations, the risk of accidents
- the suitability of the location for each type of installation
- the suitability of the climate for each type of installation.

## Original Sources of Energy

### Biomass, coal and oil

Plants use energy from sunlight to grow. Animals get their energy supplies from plants or from other animals that have eaten plants.

Coal is created from ancient plant material and oil from ancient marine animal remains. Methods of electricity generation using biomass, coal and oil (and natural gas) derive their energy indirectly from sunlight.

### Solar, wind, wave and hydroelectric

Winds are produced as a result of the uneven heating of the Earth's surface.

Waves are produced as a result of winds blowing across the surface of expanses of water.

Rainfall occurs as a result of the movement of weather systems, the mixing of air streams or the upward convection of moist air. The presence of water in lakes and rivers is, indirectly, a result of heating from the sun.

Methods of electricity generation involving winds, waves, lakes and rivers (and, of course, solar power) derive their energy from solar heating.

### Tidal and geothermal

Tides are driven by the gravitational force of the Moon. Tides extract energy from the Moon's orbit and cause it to get smaller, but this effect is negligible.

Geothermal energy is extracted from the internal energy of rocks below the surface of the Earth. This energy comes from the original formation of the Earth and from radioactive decay within the molten rocks.

Electricity generation from these two sources does not derive its energy from the Sun.

## Distribution of Electricity

However electricity is produced, it must be distributed to consumers. The National Grid system involves the use of high-voltage cables carrying large currents. Not all of the power reaches the consumer since the current heats the cables. To minimize this power loss, the current is kept small.

In order to deliver sufficient power using small currents, the voltage has to be very high (up to 400 000 V).

See page 50 for a more detailed analysis of power loss in transmission cables.

# Advantages and Disadvantages

Table 1.1 gives some advantages and disadvantages of electricity generation from different sources.

**Table 1.1**

| Source | Advantage | Disadvantage |
|---|---|---|
| Fossil fuel | Well-understood technology | Acid rain<br>Greenhouse gases<br>Depletion of scarce resources |
| Nuclear | Absence of environmental pollution when operating correctly | Danger of serious accidents<br>High fuel processing costs<br>Difficulty of safe storage and disposal of radioactive waste |
| Tidal | No fuel cost<br>No air or water pollution | Intermittent supply<br>Despoliation of habitats |
| Wave | No fuel cost<br>No air or water pollution | Lack of sites for large-scale exploitation<br>Intermittent supply |
| Hydro-electric | No fuel cost<br>No air or water pollution | Despoliation of habitats<br>Sites limited by geographical conditions |
| Wind | No fuel cost<br>No air or water pollution | Lack of sites for large-scale exploitation<br>Intermittent supply |
| Solar | No fuel cost<br>No air or water pollution | Lack of sites for large-scale exploitation<br>Intermittent supply |
| Biomass | Renewable, widely available and clean fuels | Greenhouse gases |
| Geothermal | No fuel cost<br>No air or water pollution | Sites limited by geological conditions |

## Test your understanding

1. Table 1.2 gives data for the capital and running costs of different methods of electricity generation.

   **Table 1.2**

   | Method | Capital cost per kW of generating capacity (£) | Operating/maintenance cost per kWh of electricity produced (p) | Fuel cost per kWh of electricity produced (p) |
   |---|---|---|---|
   | Coal | 870 | 0.8 | 1.3 |
   | Wind | 1300 | 0.1 | 0 |
   | Tidal | 1100 | 0.5 | 0 |
   | Geothermal | 1000 | 0.2 | 0 |
   | Nuclear | 1400 | 0.7 | 0.1 |
   | Solar | 1900 | 0.1 | 0 |

   Assess the suitability of each method of electricity generation for:
   (a) an economically developed nation
   (b) a less economically developed nation.

2. Draw energy flow diagrams to show how wind-, wave-, tidal- and fossil-fuelled electricity generation methods derive their energy from the Sun.

3. Figure 1.60 shows a skier descending the ramp of a ski jump. Figure 1.61 shows the variation of distance travelled along the ramp with time from the time the descent starts until the skier leaves the end of the ramp. The skier, of mass of 80 kg (including equipment), skis down the ramp and leaves it horizontally ($g = 9.8$ m s$^{-2}$).

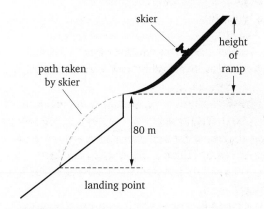

**Figure 1.60**

continued

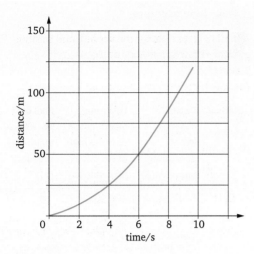

**Figure 1.61**

(a) List the energy transformations that occur as the skier skis down the ramp.
(b) Use the graph to show that the speed at which the skier leaves the ramp is about 23 m s⁻¹.
(c) Determine the skier's change of kinetic energy while descending the ramp.
(d) The skier gains 55% of the available gravitational potential energy as kinetic energy when descending the ramp. Determine the skier's change in potential energy and the height of the ramp.
(e) Figure 1.60 also shows the path taken by the skier after leaving the ramp. Assuming that there was no lift or air resistance, calculate:
  (i) the time taken for the skier to reach the ground
  (ii) the horizontal distance jumped by the skier before landing.

(See pages 32 and 33)

4 The ball-bearing in Figure 1.62 has been pushed back against a spring so that the spring is compressed by 5.0 cm. The spring constant is 120 N m⁻¹. The ball-bearing is then released.

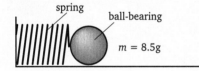

**Figure 1.62**

Calculate:
(a) the maximum force applied to the ball-bearing by the spring
(b) the maximum acceleration of the ball-bearing
(c) the maximum kinetic energy of the ball-bearing, assuming that all of the energy stored in the spring is given to the ball-bearing
(d) Suggest three reasons why the maximum kinetic energy of the ball-bearing is less than the work done compressing the spring.

5 (a) Use the electricity generation data in Table 1.2 to calculate the total cost (capital, operating and fuel costs) over the 30 year life of a coal-fired and a nuclear-powered power station.
 (b) Comment on the life-time costs of the energy from the two types of power station if decommissioning costs are also taken into account.
 (c) What are the particular problems associated with the decommissioning of nuclear power

## Work

A force can be applied without doing work provided that there is no movement in the direction of the force.

Work is done whenever a force is applied and there is movement that has a component in the direction of the force:

$$\text{work done} = Fs$$

where $F$ = force used and $s$ = distance moved in the direction of the force.

Energy is transformed from one sort to another whenever work is done. The amount of energy transformed is equal to the amount of work done.

Work has the same unit as energy, i.e. the joule (J). The joule is defined as the work done when a force of 1 N moves by a distance of 1 m:

$$1 \text{ joule} = 1 \text{ newton metre}$$
$$1 \text{ J} = 1 \text{ N m}$$

## Gravitational Potential Energy

To lift a body, the force ($F$) needed is equal and opposite to the weight of the body.

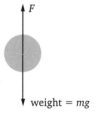

**Figure 1.63**

When the body is lifted up, work is done against gravity. The body gains **gravitational potential energy**. The potential energy gained is equal to the work done:

$$\text{potential energy} = F\Delta h$$
$$= mg\Delta h$$

# Kinetic Energy

When a force accelerates a body, work is done and the body gains **kinetic energy**. The amount of kinetic energy the body gains is equal to the work done.

Consider a body accelerated horizontally from rest. There is no change in potential energy so the body only gains kinetic energy:

$$\text{work done} = \text{force} \times \text{distance moved}$$

$$\text{kinetic energy} = Fs = mas, \text{ since } F = ma$$

but $$v^2 = u^2 + 2as$$

$$v^2 = 2as, \text{ as } u = 0$$

or $$as = \tfrac{1}{2}v^2$$

so $$\text{kinetic energy} = m(\tfrac{1}{2}v^2)$$

or $$\text{kinetic energy} = \tfrac{1}{2}mv^2$$

### Kinetic energy of falling bodies

When a body falls, it accelerates. As it gets faster, its kinetic energy increases and its potential energy decreases:

$$\text{gain in kinetic energy} = \text{loss in potential energy}$$

When a body falls from rest (i.e. has no initial velocity):

$$\tfrac{1}{2}mv^2 = mg\Delta h$$

This approach can be used as an alternative to the equations of motion to find the velocity of a falling body.

**EXAMPLE:** An object is dropped 15 m from rest. Find its velocity when it has fallen 15 m ($g = 9.8 \text{ m s}^{-2}$).

$$\tfrac{1}{2}mv^2 = mg\Delta h$$
$$\tfrac{1}{2}v^2 = g\Delta h$$
$$v^2 = 2g\Delta h$$
$$= 2 \times 9.8 \times 15$$
$$= 294$$
$$v = 17 \text{ m s}^{-1}$$

### Test your understanding

1. A 0.15 kg ball is dropped vertically and hits the ground at a speed of 30 m s⁻¹ ($g = 9.8 \text{ m s}^{-2}$). Calculate:
   (a) the kinetic energy of the ball immediately before it hits the ground
   (b) the initial potential energy of the ball
   (c) the height from which the ball was dropped.

2. Water goes over a waterfall at the rate of 35 000 kg s⁻¹. The vertical speed of water just as it reaches the bottom of the water fall is 12 m s⁻¹ ($g = 9.8 \text{ m s}^{-2}$). Calculate:
   (a) the height of the waterfall
   (b) the rate of energy dissipation when the water gets to the bottom (assume that the water has no horizontal component of velocity).

3. A skier with a mass of 85 kg skis down a ski jump that has a vertical height of 40 m. Assuming that friction is negligible, calculate the skier's speed at the bottom of the ramp ($g = 9.8 \text{ m s}^{-2}$).

4. (a) A ball is dropped vertically from a height of 2.0 m above the ground. Each time the ball bounces, it loses 40% of its kinetic energy. Calculate:
   (i) the velocity of the ball just before it hits the ground for the first time
   (ii) the velocity of the ball immediately after one impact
   (b) Sketch a velocity–time graph covering the time from when the ball is dropped until after the fourth impact. Assume that the duration of each impact is zero ($g = 9.8 \text{ m s}^{-2}$).

# Forces and Fluids

When water flows along a pipe, it can do so in one of two ways:
- a streamline or laminar flow in which the particles move parallel to the pipe walls (Figure 1.64)
- a turbulent flow in which the movement of particles is irregular and contains eddies or whirlpools (Figure 1.65).

**Figure 1.64**

**Figure 1.65**

Slow-moving water is more likely than fast-moving water to exhibit laminar flow. Other fluids (liquids and gases) also exhibit the same types of flow patterns. Viscous fluids are more likely to have laminar flow.

## Lift Forces

When fluids move quickly, their pressure drops. This can be illustrated by blowing across the top surface of a dangling sheet of paper (Figure 1.66).

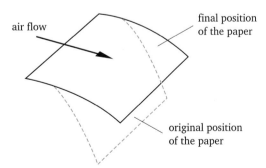

**Figure 1.66**

The air moves quickly over the top surface so the pressure is low. The air is stationary below the paper so the pressure stays high. The pressure difference causes the paper to lift (Figure 1.67).

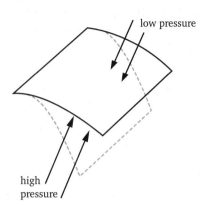

**Figure 1.67**

This phenomenon, known as the Bernoulli principle, is used in the design of aircraft wings (Figure 1.68).

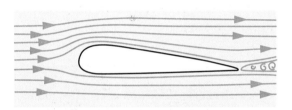

**Figure 1.68**

Air moves over the top and bottom surfaces of the wing at the same time. Because of the curved shape of the wing, the distance moved by the air is greater over the top surface. The air travels faster over the top surface of the wing.

The pressure is reduced more at the top of the wing than at the bottom and the pressure is bigger at the bottom of the wing than it is at the top. The wing is therefore pushed upwards (Figure 1.69).

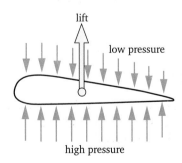

**Figure 1.69**

# Newton's Laws of Motion

It is not necessary to memorize these laws but it is necessary to understand what they tell us about the movement of objects.

**First law**: A body will remain at rest, or in uniform motion in a straight line, unless acted on by an external force.

Bodies only *change* the way in which they move in response to an external force. Forces can change a body's velocity by changing its direction or its speed.

Bodies do not need to have a force applied to them to keep them moving at a steady speed in a straight line.

When it seems as though a force is needed to maintain steady motion, this is usually because of forces resisting motion. The cyclist in Figure 1.79 is moving with constant velocity.

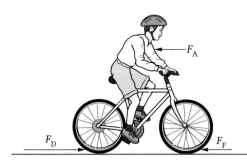

**Figure 1.79**

The driving force ($F_D$) is equal and opposite to the sum of the forces of friction ($F_F$) and air resistance ($F_A$) that oppose the motion. The net force is zero. The bicycle is in **equilibrium**.

## Equilibrium

When a body is in equilibrium, it is either stationary or moving with constant speed in a straight line. A body is in equilibrium when:
- the sum of the moments about any point is zero (see page 20)
- the resultant force on the body is zero.

**Second law**: When a body is acted on by an external force, its acceleration is proportional to the applied force or

$$F = ma$$

where $F$ = the applied force, $m$ = the mass of the body and $a$ = the acceleration produced.

The SI unit of force is derived from this equation: **a force of 1 newton, when applied to a body of mass 1 kg, will cause an acceleration of 1 m s$^{-2}$.**

## Weight and Mass

Weight is the force, caused by gravity, that pulls the body towards the centre of the Earth (or towards whichever planet it is on). Weight, $W$, is affected by the strength of the gravitational field:

$$W = mg$$

where $m$ = the mass of the body and $g$ = gravitational field strength.

The mass of the body is unaffected by the strength of the gravitational field.

## Acceleration Due to Gravity

All objects accelerate at the same rate (approximately 9.8 m s$^{-2}$) when falling freely under the influence of gravity near the Earth's surface. Figure 1.80 shows two bodies accelerating freely under gravity. Their accelerations are caused by their weights.

**Figure 1.80**

For the larger body:
$$a_1 = \frac{F}{m}$$
$$a_1 = \frac{m_1 g}{m_1}$$
$$a_1 = g$$

For the smaller body:
$$a_2 = \frac{F}{m}$$
$$a_2 = \frac{m_2 g}{m_2}$$
$$a_2 = g$$

Both bodies accelerate downwards at a rate (in m s$^{-2}$) that is numerically equal to the strength of the gravitational field (in N kg$^{-1}$).

**Third law**: Every action has an equal and opposite reaction.

The chair in Figure 1.81 has weight $W$ caused by gravitational attraction from the Earth. The chair attracts the Earth with an equal and opposite force $F_E$.

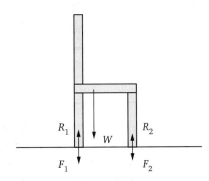

**Figure 1.81**

$W$ and $F_E$ form an action–reaction pair.

The force applied by the chair leg, $F_1$, is equal and opposite to the force of the floor supporting the chair, $R_1$. $F_1$ and $R_1$ form another action–reaction pair, as do $F_2$ and $R_2$.

The chair is in equilibrium as

$$W = R_1 + R_2$$

### Test your understanding

1. Decide whether or not the following bodies are in equilibrium. If they are not, calculate the resultant force and state the magnitude and direction of their accelerations.

   (a)

   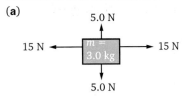

   **Figure 1.82**

   (b)

   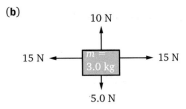

   **Figure 1.83**

   (c)

   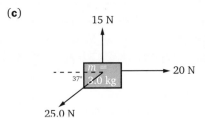

   **Figure 1.84**

   (d)

   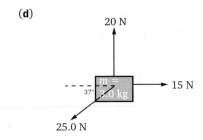

   **Figure 1.85**

2. The following bodies are held on the inclined planes by friction. By resolving the forces, find the magnitudes of the normal reactions and the frictional forces. Indicate the directions of the forces on copies of the diagrams.

   (a)

   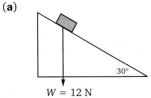

   **Figure 1.86**

   (b)

   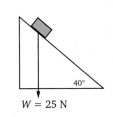

   **Figure 1.87**

3. A car has a mass of 750 kg. It is on a road which has a gradient of 30°. The maximum frictional force between the car and the road is 3500 N.
   (a) Draw a diagram of the road and the car showing the magnitude of the following:
      (i) the weight of the car
      (ii) the frictional force on the car
      (iii) the component of the car's weight that is acting parallel to the road
      (iv) the component of the car's weight that is acting perpendicular to the road
      (v) the normal reaction of the road on the car.
   (b) Decide whether or not the car will be in equilibrium and if not what its acceleration will be.

# Mechanical Oscillations

An oscillation is a movement in which a body moves repeatedly backwards and forwards past a fixed point. The movement of a pendulum is an example of an oscillation.

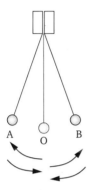

**Figure 1.88**

A complete oscillation occurs when the pendulum bob moves from O to A, back past O to B and back to O again. Alternatively, a movement from A to B and back again would be one complete oscillation.

The **period** of the pendulum is the time taken for one complete oscillation.

The **amplitude** of the oscillation is the maximum distance moved by the pendulum bob from the mean position O.

When timing oscillations, errors can be minimized by timing a large number, $n$, of oscillations and dividing the total time by $n$.

When counting oscillations in order to find the period, it is wise to use a **fiducial** mark at the mean position O (Figure 1.89).

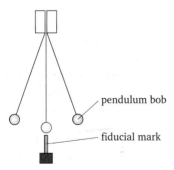

**Figure 1.89**

**Note that for AS you are not required to recall or apply the formulae in the examples that follow.**

You may be required to suggest how factors in the design of an oscillator might affect its period and be able to test experimentally the validity of equations that you are given and derive results from your data.

## Finding the Acceleration Due to Gravity, g

The period of the oscillation depends only on the length, $l$, of the pendulum and on the gravitational field strength, g. It is not affected by the amplitude of the oscillation, provided that it is kept reasonably small, nor is it affected by the mass of the bob.

The period, $T$, is given by:

$$T = 2\pi\sqrt{\frac{l}{g}}$$

Squaring each side of the equation gives:

$$T^2 = 4\pi^2 \frac{l}{g}$$

To find $g$:
- measure $T$ for several values of $l$ over a wide range
- for each value of $T$ calculate the value of $T^2$
- plot a graph with $T^2$ on the $y$-axis and $l$ on the $x$-axis (Figure 1.90)
- measure the gradient $m$ of the graph.

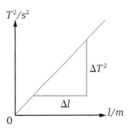

**Figure 1.90**

$$m = \frac{\Delta T^2}{\Delta l}$$

also

$$m = \frac{4\pi^2}{g}$$

The acceleration due to gravity, $g$, can be found from these two equations.

## Mass/Spring Oscillators

Similar oscillations can be observed in a mass/spring system (Figure 1.91).

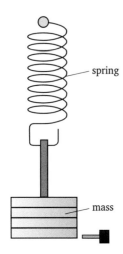

**Figure 1.91**

The period of the oscillation, $T$, is given by:

$$T = 2\pi\sqrt{\frac{m}{k}}$$

$$T = 2\pi\sqrt{\frac{\Delta l}{g}}$$

The second of these equations may be used to find $g$ as follows:
- add a load on to the spring and note the spring extension, $\Delta l$
- measure the time period, $T$, of the oscillations caused when the load is pulled down a little and released. Repeat for a range of values of $\Delta l$
- plot a graph of $T^2$ ($y$-axis) against $\Delta l$ ($x$-axis)
- find the gradient of the graph:

$$\text{gradient} = 4\pi^2/g$$

## Damping of Oscillations

If either of the systems described is set to oscillate, the amplitude will reduce and, in the end, the system will come to rest. This is because the system loses energy as work is done against friction or air resistance. The reduction in amplitude is called **damping** and is shown in Figure 1.92.

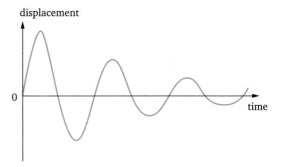

**Figure 1.92**

Notice that the amplitude of the oscillation reduces with time but the period, and hence the frequency of the oscillation, stay constant.

## Using Data Logging to Investigate Oscillations

Figure 1.93 shows a pendulum oscillating with its bob in a bath of copper sulphate solution.

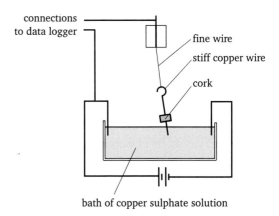

**Figure 1.93**

The pendulum is made of fine conducting wire. As it oscillates, the voltage across the data logger leads varies (the copper sulphate bath acts as a potential divider).

The voltage across the leads is approximately proportional to the distance between the pendulum bob and the left-hand side of the bath.

The data logger records the varying voltage. A printout of the voltage variation enables you to deduce the period of the oscillation and observe the variation of amplitude as the motion is damped.

## Test your understanding

1. The data in Table 1.3 relate to the oscillation of a mass/spring system.

   **Table 1.3**

   | Extension, $\Delta l /$ $10^{-3}$ m | Time for 20 oscillations/s | | | Periodic time, $T$/s | $T^2/s^2$ |
   |---|---|---|---|---|---|
   | | 1st time | 2nd time | Average | | |
   | 17 | 5.1 | 5.5 | | | |
   | 34 | 7.4 | 7.6 | | | |
   | 51 | 9.1 | 9.0 | | | |
   | 68 | 10.4 | 10.7 | | | |
   | 84 | 11.6 | 11.4 | | | |
   | 101 | 12.9 | 12.7 | | | |
   | 118 | 13.9 | 13.8 | | | |

   (a) Complete the table and use the data to plot a graph of $T^2$ (on the y-axis) against $\Delta l$ (on the x-axis).

   (b) Measure the gradient of the graph and use your answer to determine the acceleration due to gravity, $g$.

   (c) Suggest which of the results from the table are the least accurate and explain why.

# Electricity

## Current (I)

Current is the movement of charged particles. In metals, the moving charged particles are **electrons**.

Metals are good electrical conductors because they have many electrons that are free to move.

About one electron from every atom is a 'free electron'. The atoms that lose these electrons have fewer than their normal complement of electrons and are called **lattice ions**.

When a voltage is applied to a metal conductor, the free electrons are attracted to the positive terminal and repelled by the negative terminal (Figure 1.94).

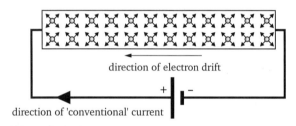

**Figure 1.94**

In the early days of the study of electric currents it was assumed that the moving particles carried a positive charge. It became accepted to show the current direction from positive to negative. When it was realized that electrons were negative, it was too difficult to change.

## The Ampere (amp)

It would be possible to measure currents in electrons per second but this would be an inconveniently small unit.

Instead, the unit for current is the **ampere** (amp). A current of 1 A is equivalent to the movement of **1 coulomb** (1 C) of charge past a point in 1 second.

1 C is equivalent to the sum of the charge on approximately 6.25 million million million electrons ($6.25 \times 10^{18}$ electrons).

## The Relationship between Current and Charge

Current = rate of flow of charge

$$I = \frac{\Delta Q}{\Delta t}$$

where $\Delta Q$ = charge and $\Delta t$ = time.

## Movement of Electrons and Lattice Ions

Lattice ions vibrate. At higher temperatures they vibrate faster and with bigger amplitudes. They collide with free electrons which, as a result, move randomly around the lattice, even when there is no current.

When there is a current, the electrons have two parts to their movement:
- random movement caused by collisions with lattice ions
- drift towards the positive terminal of the supply.

The random movement of the electrons is fast compared with their drift velocity.

## Current, Charge and Drift Velocity

If the number of free electrons for a given type of wire is known, the drift velocity can be calculated. Figure 1.95 shows a section, PQ, of wire carrying a current I. The cross-sectional area of the wire is A.

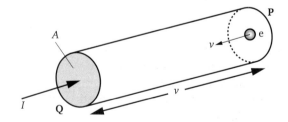

**Figure 1.95**

Consider what happens to an electron, e, starting at P and moving at $v$ m s$^{-1}$. In 1 second, the electron will move $v$ metres, i.e. from P to Q.

volume of wire in section PQ = $Av$

If there are $n$ free electrons per m³ in the wire:

$$\text{total number of electrons in PQ} = nAv$$

Each electron has a charge of $e$, so:

$$\text{total free charge in PQ} = nAve$$

Since in the second we are considering the electron shown moves from P to Q, all of the electrons that were in PQ to start with will move past Q in that second:

$$\text{charge passing Q in 1 s} = I = nAve$$

The derivation of this equation need not be remembered but the equation itself is important.

In some current situations the charge carriers may not be electrons. A more general form of the equation, for particles of charge $q$, is:

$$I = nAvq$$

## Conduction in Liquids and Gases

Liquids and gases will conduct electricity if there are charged particles present that are free to move.

In ionic solutions, positive and negative ions are available as charge carriers (Figure 1.96).

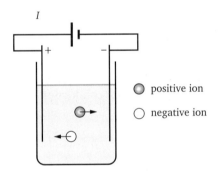

**Figure 1.96**

Gases do not tend to have charge carriers available so they are insulators.

Charge carriers can be created by ionizing atoms.

Electrical discharges such as lightning or a spark from a Van de Graaff generator happen when some of the air becomes ionized:
- a very small number of charged particles is present in the air
- if a high voltage is applied, these few particles are accelerated to high speeds
- collisions with air molecules result in further ionization and each ion produces more charge carriers in a cascade
- when an ionization occurs, the electron which has been removed and the remaining positive ion become charge carriers in the discharge.

Test your understanding

1. Calculate the current flowing when:
   (a) 12 C moves past a point in 2.0 s
   (b) 0.58 C moves past a point in 120 s.
2. Calculate the charge which has passed a point in a circuit when there is a current of:
   (a) 4.0 A for 60 s
   (b) 12 mA for 25 s.
3. Calculate how long it takes for the following charges to pass a point in a circuit:
   (a) 6.0 C, when the current is 0.50 A
   (b) 48 C, when the current is 36 mA.
4. Calculate the number of electrons passing a point in the circuits referred to in question 2.
5. Calculate the drift velocity of electrons along a wire of diameter 0.32 mm, carrying a current of 2.4 A.

$$n = 8.5 \times 10^{28} \text{ m}^{-3}$$
$$e = 1.6 \times 10^{-19} \text{ C}$$

Suggested Investigations

(a) The variation of current with applied voltage for an ionic solution.
(b) The variation of current with concentration for an ionic solution.
(c) The variation of current with electrode size or separation for an ionic solution.

## Resistance ($R$)

A component's resistance is defined as the ratio of the voltage across it to the current flowing through it.

$$R = \frac{V}{I}$$

The unit of resistance is the **ohm** ($\Omega$). A component has a resistance of 1 $\Omega$ if the current flowing through it is 1 A when the voltage between its ends is 1 V.

For metallic conductors, the resistance is constant, providing that the temperature of the conductor stays the same. This means that the graph of current against voltage is a straight line (Figure 1.97).

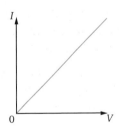

**Figure 1.97**

## Resistance in Metals

Electrons collide with lattice ions. Each collision changes the electron's velocity and impedes its drift. More energy from the power supply is needed to accelerate the electron again.

## Variation of Resistance with Temperature

At low temperatures:
- lattice ions have small amplitude and low frequency vibrations
- electrons are quite likely to pass the ions without interference
- the resistance is low.

At higher temperatures:
- lattice ions vibrate faster and with greater amplitude
- electrons are more likely to collide with ions
- the resistance of the conductor is higher.

**Ohm's law** takes account of the variation of resistance with temperature.

**Ohm's law**: the current through a metallic conductor is proportional to the voltage across its ends, providing the temperature remains constant.

The equation for Ohm's law is most often given in the form:

$$V = IR$$

## The Heating Effect of Currents

When there is no current, the free electrons are in equilibrium with the lattice ions. Collisions between electrons and lattice ions still happen. In some collisions, electrons gain kinetic energy from the lattice ions. In other collisions, the lattice ion gains energy from the electrons.

When there is a current:
- the electrons have been accelerated using electrical energy from the power supply
- the electrons have more kinetic energy than normal
- when an electron collides with a lattice ion, the electron is more likely to lose some of its energy to the lattice ion
- the ion gains vibrational kinetic energy. Increasing the energy of the lattice is equivalent to heating it
- electrical energy from the power supply has been converted into internal energy in the conductor.

The energy change in each collision is small but many billions of collisions occur. The conductor subsequently loses some of its energy by conduction, convection or radiation. This happens whenever a current flows through any component that has resistance.

Test your understanding

1. Calculate the voltages and resistances that are missing from Table 1.4.

**Table 1.4**

| V | I | R |
|---|---|---|
| 12 V | 4.0 A | |
| 230 V | 18 mA | |
| | 2.5 A | 180 Ω |
| | 3.4 mA | 27 kΩ |
| 230 V | 0.25 A | |
| $4.8 \times 10^6$ V | 24 mA | |

## Factors Affecting the Resistance of Wires

The resistance of a wire is proportional to its length. Electrons passing through a long wire will have more collisions than when they pass through a short wire:

$$R \propto l$$

The resistance of a thick wire is less than the resistance of a thin wire. Thick wires have more free electrons available for conduction. Resistance is *inversely* proportional to cross-sectional area:

$$R \propto \frac{1}{A}$$

Some materials are good conductors and some are poor conductors. Good conductors have more free electrons available for conduction than poor conductors.

Plastics and ceramics are insulators because they have almost no free electrons. Metals are extremely good conductors because they have so many free electrons.

Resistance is proportional to the resistivity, $\rho$, of the material from which the conductor is made:

$$R \propto \rho$$

Combining these relationships gives:

$$R = \frac{\rho l}{A}$$

Rearranging this equation gives:

$$\rho = \frac{RA}{l}$$

A sample of material with length 1 m and cross-sectional area 1 m² will have a resistance that is numerically equal to its resistivity.

The unit of resistivity is Ω m.

### Table 1.5 Resistivities

| Material | Resistivity (Ω m) |
|---|---|
| Graphite | $1.4 \times 10^{-5}$ |
| Copper | $1.6 \times 10^{-8}$ |
| Constantan | $4.9 \times 10^{-7}$ |
| Nichrome | $1.1 \times 10^{-6}$ |
| Steel | $1.8 \times 10^{-7}$ |
| Glass | $\sim 10^8$ |

## Resistance of Other Components

### Filament lamp

The current in a filament lamp heats the filament so that it glows. A larger current causes an increase in temperature. The increase in temperature causes an increase in resistance (Figure 1.98).

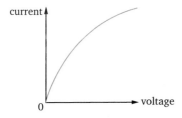

**Figure 1.98**

### Thermistors and light-dependent resistors

These are made from semiconductors, which are materials that have few charge carriers available because the electrons are bound to their atoms unless stimulated in some way.

### Thermistors

Heating the material will give the electrons enough energy to free themselves from their atoms and become available for conduction.

Thermistors have large changes of resistance when their temperature changes. They are particularly useful as temperature sensors (Figure 1.99).

Resistors made from metal wire also suffer changes of resistance with temperature so they too can be used as temperature sensors. However, the change of resistance with temperature for metals is not so large.

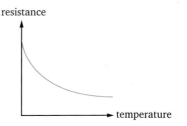

**Figure 1.99**

### Light-dependent resistor (LDR)

Light energy falling on the surface of an LDR liberates electrons from their atoms and makes them available for conduction.

Increasing the light level increases the number of electrons and decreases the resistance (Figure 1.100). Because of this change in resistance LDRs can be used as light sensors.

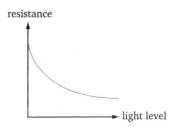

**Figure 1.100**

### Diodes

Diodes have two different types of semiconductor arranged so that conduction is possible when the voltage is applied in one direction and not in the other.

When the voltage is the right way round, a small voltage, around 0.6 V, needs to be applied before the diode will conduct at all. When the applied voltage is above this level, the resistance of the diode falls to a very low level (Figure 1.101). When the voltage is applied in the opposite direction, the diode will not conduct at all. This makes the diode ideal for converting alternating current into direct current (Figure 1.102).

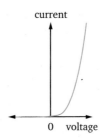

**Figure 1.101**

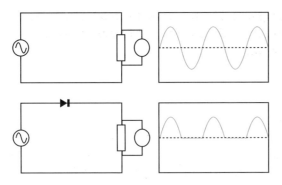

**Figure 1.102**

In the top diagram, the oscilloscope trace shows alternating current. In the lower diagram, the diode has prevented the current from flowing in one direction. The oscilloscope shows intermittent, unsmoothed direct current.

### Test your understanding

1. Using the values of resistivity given on page 45, calculate the resistances of the following conductors:
   (a) a copper wire of length 2.3 m and cross-sectional area $1.5 \times 10^{-9}$ m$^2$
   (b) a circular cross-sectioned glass rod of length 24 cm and radius 4.0 mm. (Remember to convert your dimensions into metres before doing the calculation.)
2. A graphite block has length 12 mm, width 4.0 mm and height 5.0 mm. Current can be passed through it between opposite faces. Calculate the three values of resistance that it can have.
3. Calculate the length of 0.32 mm diameter nichrome wire that has the same resistance as a 3.4 m length of 0.26 mm diameter constantan wire.

## Superconductivity

When a metallic resistor is cooled, its resistance falls. The rate of change of resistance with temperature suggests that the resistance will become zero at around absolute zero ($-273$ °C) (Figure 1.103).

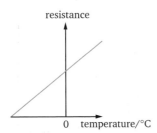

**Figure 1.103**

In fact, when metals are cooled it is found that their resistance reaches zero at temperatures of a few kelvin. Materials with no electrical resistance are called **superconductors**.

Because the resistance is zero, it is possible to maintain a current in a very low temperature coil without having a voltage driving it.

Superconductors are potentially useful in any application where large currents are used. Without resistance, there are no energy losses.

Applications of superconductors include:
- power cables
- monorail trains
- generators, transformers and motors
- body scanners such as magnetic resonance imaging scanners, which need strong magnetic fields produced by large currents
- electromagnets for particle accelerators.

In recent years, materials have been developed that will superconduct at much higher temperatures. For example, a magnesium–barium–calcium–copper oxide material has been developed that exhibits superconductivity at 133 K ($-140$ °C). The temperature at which the material becomes a superconductor is called its **transition temperature** (Figure 1.104).

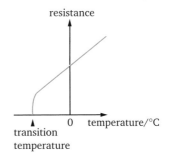

**Figure 1.104**

Superconducting cables are expensive because of the need to keep them refrigerated at low temperatures. Figure 1.105 shows a superconducting cable.

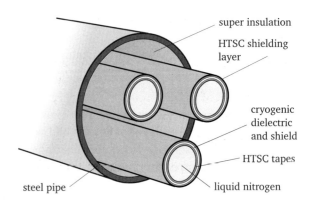

**Figure 1.105**

## Voltage and Electromotive Force (emf)

Voltage can be regarded as electrical 'force' that pushes charge round a circuit. However, it is more useful to think of it in terms of energy changes.

To move a charge through a resistor, work has to be done on it. In a simple circuit, the energy to do this work comes from the chemical energy in the combination of materials that make up a cell or battery. The work done by the cell, $W$, for each coulomb of charge moved around the circuit is called the **electromotive force (emf)**, $E$, of the cell:

$$W = Eq$$

where $q$ = the amount of charge moved around the circuit, or

$$E = \frac{W}{q}$$

> **Key point**
> 
> Emf refers to the work done (per unit charge) by the power supply, e.g. the cell, battery or generator. Use the term emf when electrons are being given the energy to do work.

## Potential Difference

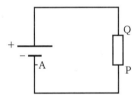

**Figure 1.106**

An electron at position A in Figure 1.106 is repelled by the negative terminal and attracted to the positive terminal. As it moves round the circuit, the electron does work, raising the temperature of the resistor. It has potential energy before it passes through the resistor.

When the electron has passed the resistor, it has lost its energy.

There is a **potential difference** between P and Q. The potential difference, usually given the symbol $V$, is the amount of energy produced for every coulomb of charge that passes.

energy transformed in the resistor = $Vq$

where $q$ = the amount of charge moved through the resistor.

> **Key point**
> 
> Potential difference refers to the voltage across components where the electrons are giving up their energy.

In Figure 1.106, the emf of the cell is equal and opposite to the potential difference across the resistor.

## Potential Differences in Series Circuits

In Figure 1.107 the cell's emf is 1.5 V. This means that every coulomb of charge that moves round the circuit must result in 1.5 J of energy being transformed from chemical energy in the cell into internal energy in the resistors.

Some of the energy is transformed in $R_1$ and the rest is transformed in $R_2$. More work is done passing the charge through the larger resistor, $R_1$, and so the change in potential is bigger across $R_1$.

Since the emf is 1.5 V, the sum of the potential differences across the resistors must also be 1.5 V. The 1.5 V is shared between the resistors in proportion to their resistances.

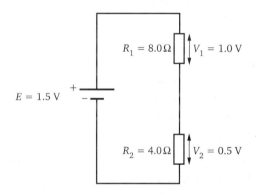

**Figure 1.107**

> **Key point**
> 
> The sum of the emfs in the loop of the circuit is balanced by the sum of the potential differences across the other components.

## Potential Difference across Components in Parallel

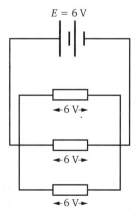

**Figure 1.108**

In the circuit in Figure 1.108 three different resistors are connected in parallel. The wires connecting the resistors to the battery have no resistance.

The electrons lose no potential energy as the charge moves through the wires. The only change in the charge's potential energy occurs when it moves through the resistors. It is therefore only the resistors that have a potential difference across them.

Since all the resistors are connected identically to the battery, they all have the same potential difference across them. In this case the potential difference across each resistor is the same as the emf of the battery.

### Key point
The potential difference is the same for all components connected in parallel.

## Current at Junctions in a Circuit

Charge is conserved in any circuit: it is not created or destroyed and it does not leak out of the system. The sum of the currents entering a junction must therefore be the same as the sum of the currents leaving the junction (Figure 1.109).

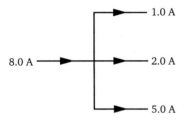

**Figure 1.109**

## Resistors in Series

Figure 1.110 shows two circuits carrying the same current and having the same emf.

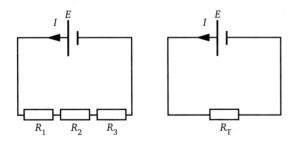

**Figure 1.110**

Because all of the other conditions are identical, the resistor ($R_T$) in the second circuit must be the equivalent of the series combination ($R_1$, $R_2$ and $R_3$) in the first circuit.

The sum of the potential differences across $R_1$, $R_2$ and $R_3$ is equal to $E$. The potential difference across $R_T$ must also be equal to $E$:

$$E = IR_1 + IR_2 + IR_3$$

and

$$E = IR_T$$

so

$$IR_1 + IR_2 + IR_3 = IR_T$$

or

$$I(R_1 + R_2 + R_3) = IR_T$$

The resistance of a series combination of resistors is therefore:

$$R_1 + R_2 + R_3 = R_T$$

## Resistors in Parallel

The equation for the equivalent resistance ($R_T$) of three resistors in parallel is:

$$\frac{1}{R_1} + \frac{1}{R_2} + \frac{1}{R_3} = \frac{1}{R_T}$$

### Test your understanding

1. Draw two equivalent circuits, similar to Figure 1.110 but showing three resistors in parallel and the equivalent resistor $R_T$.
   By comparing the sum of the currents through $R_1$, $R_2$ and $R_3$, prove the equation for the resistance of a parallel combination of resistors.

## Emf and Internal Resistance

All electrical power sources have a built-in or internal resistance. This is an unintentional consequence of how the device is made:
- cells are made of combinations of materials through which the charge must pass
- generators have coils of wire through which the charge must pass.

The effects of internal resistance are to limit:
- the current that can be drawn from the device
- the potential difference across the terminals of the device.

The cell in Figure 1.111 has an internal resistance $r$ of 0.5 Ω. It is shown on the diagram as being an internal resistance as it is positioned between the cell terminals P and Q.

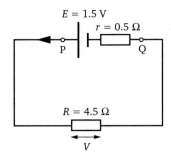

**Figure 1.111**

The internal resistance increases the total resistance of the circuit:

$$R_T = R + r$$
$$= 5.0\ \Omega$$

The current in the circuit is less than it would have been without an internal resistance:

$$I = \frac{E}{R_T}$$
$$= 0.30\ A$$

The potential difference across R is:

$$V = IR$$
$$= 0.30 \times 4.5$$
$$= 1.35\ V$$

The potential difference across the cell terminals PQ is the same as V. The potential difference PQ is less than the full emf of the cell (1.5 V). The full emf of the cell is not available to the external circuit because the cell has done some of its work moving the charge through its own internal resistance.

The 'lost voltage' that is not available to the external circuit is $Ir$. In this case it is 0.15 V.

The difference between the emf and the cell terminal potential difference is equal to the voltage 'lost' across $r$:

$$V = E - Ir$$

## Significance of Internal Resistance

If a power supply has an internal resistance, the current that can be drawn from it is limited. The maximum current is drawn from a supply when it is short circuited.

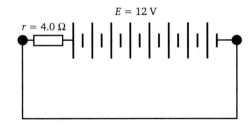

**Figure 1.112**

Figure 1.112 shows eight 1.5 V cells in series. Each has an internal resistance of 0.5 Ω. The cells have been short circuited. The total resistance in the circuit is 4.0 Ω, giving a current I of:

$$I = \frac{12}{4}$$
$$= 3.0\ A$$

These cells could not be used to power a car's starter motor. Although the voltage is correct, a starter motor needs a much higher current, around 150 A.

High-voltage power supplies are often made with a very large internal resistance as a safety precaution.

## Energy and Power in Circuits

When a charge of $q$ moves through a potential difference of V, the energy transformed is given by:

$$\text{energy} = Vq$$

Since $q = It$, this energy equation can be rewritten as:

$$\text{energy} = VIt$$

By using the Ohm's law equation this can also be rearranged into other useful forms:

$$\text{energy} = VIt,\ \text{but}\ I = \frac{V}{R}$$

so

$$\text{energy} = V\left(\frac{V}{R}\right)t$$

$$= \frac{V^2}{R}t$$

or

$$\text{energy} = VIt,\ \text{but}\ V = IR$$

so

$$\text{energy} = IRIt$$
$$= I^2Rt$$

Since power is work done (or energy used) per second:

$$\text{power} = \frac{\text{energy}}{\text{time}}$$

These energy equations can be changed into equations for power:

$$\text{energy} = VIt \quad \text{gives} \quad \text{power} = VI$$

$$\text{energy} = \frac{V^2}{R}t \quad \text{gives} \quad \text{power} = \frac{V^2}{R}$$

$$\text{energy} = I^2Rt \quad \text{gives} \quad \text{power} = I^2R$$

## Power Loss in Electricity Cables

In the National Grid system, a lot of electrical power is delivered over large distances. Since $P = VI$, either the voltage must be very high or the current must be very high.

All power cables have some resistance. Since the power loss in a resistor is $I^2R$, high currents lead to large power losses. To reduce power losses, electricity is delivered at voltages of 400 000 V.

Consider a power station sending 50 MW to a town 10 km away, involving 20 km of cable. A 1 km long power cable has a resistance of approximately 0.063 Ω so the total resistance of the cable is 1.26 Ω.

$$I = \frac{P}{V}$$
$$= \frac{50\,000\,000}{400\,000}$$
$$= 125 \text{ A}$$

$$\text{Power loss} = I^2R$$
$$= 125^2 \times 1.26$$
$$= 20\,000 \text{ W}$$

This is only 0.04% of the power sent from the power station.

Now consider the voltage loss across the power cable:

$$V = IR$$
$$= 125 \times 1.26$$
$$= 160 \text{ V}$$

This is only 0.04 % of the delivery voltage.

Power loss also affects electrical wires used for communications, for example telephone cables. Signal quality is lost because of the resistance of the wires. The received voltage is less than the voltage of the transmitted signal. The voltage drop across the cable can be determined in a similar way to that for power transmission cables.

## Alternating Current for Mains Power

The need for high voltages for electricity transmission means that we have to use alternating current instead of direct current. These high voltages are unsuitable for domestic and industrial consumption.

Transformers can change electricity to a higher voltage for distribution and back to a lower voltage for consumption. Transformers can only be used with alternating current.

The following sequence applies to the distribution of electricity:
- Power stations produce electricity at 33 kV.
- Transformers increase the voltage to 440 kV for long-distance distribution.
- Substations reduce the voltage to 132 kV for large individual consumers and for local distribution.
- Local substations reduce the voltage to 230 V for distribution to individual houses.

**Test your understanding**

1. (a) For the circuit shown in Figure 1.111 calculate:
   (i) the current
   (ii) the potential difference across the internal resistor
   (iii) the potential difference across the cell terminals.
   (b) Repeat these calculations for values of $R$ of:
   (i) 99.5 Ω
   (ii) 19.5 Ω
   (iii) 9.5 Ω
   (iv) 0.5 Ω
   (v) 0 Ω (i.e. when the terminals of the cell are short circuited).
2. Calculate the terminal potential difference in the circuit in Figure 1.113.

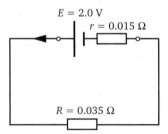

**Figure 1.113**

3. A car battery delivers 150 A to a starter motor of resistance 0.090 Ω. The emf of the battery is 14 V. Calculate:
   (a) the internal resistance of the battery
   (b) the terminal potential difference when the battery is supplying a current of 150 A.
4. A laboratory power supply with a maximum voltage of 2000 V is built with a resistance of $5 \times 10^6$ Ω. Explain why.
5. The power delivery system described in question 4 is used to deliver the same power through the same cables but using a voltage of 12 000 V. Calculate:
   (a) the power loss
   (b) the percentage power loss
   (c) the voltage drop across the cables
   (d) the percentage of the original voltage that this represents.

# Potential Dividers

Resistors in series share the applied voltage between them.

The arrangement in Figure 1.114 can be used to select a required potential difference.

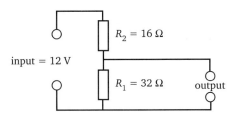

**Figure 1.114**

By working out the total resistance of the circuit it can be seen that the current is 0.25 A. Using $V = IR$ for each resistor, it is clear that the potential difference across $R_2$ is 4.0 V and the potential difference across $R_1$ is 8.0 V.

By connecting the output across $R_1$, the chain of resistors has been used to create an 8.0 V supply from the 12 V supply. Since the supply potential difference has been divided into two parts, this system is called a **potential divider**.

The voltage is divided between the two resistors in proportion to their resistances. $R_1$ is two-thirds of the total resistance so the potential difference across $R_1$ is two-thirds of the total potential difference.

For the circuit in Figure 1.114 the output voltage can be calculated using the formula

$$\text{output voltage} = \left(\frac{R_1}{R_1 + R_2}\right) \times \text{input voltage}$$

## Using Potential Dividers

Temperature and light sensors can be used in potential divider circuits to activate alarms or control circuits (Figure 1.115).

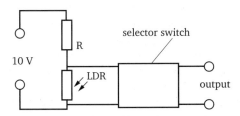

**Figure 1.115**

The 10 V supply voltage is shared between the LDR and R in proportion to their resistances.

The selector switch gives an output when it detects a voltage of more than 5.0 V. This output may be used to control an automatic light or it can be connected to a data logger.

When light levels are high, the resistance of the LDR is low so a small share of the 12 V is detected by the selector switch. As it gets darker, the resistance of the LDR increases. When the resistance of the LDR increases to be as large as the resistance of R, the voltage across the LDR reaches 5.0 V and the selector switch is operated.

The sensitivity of the circuit can be changed by using a variable resistor in place of R (Figure 1.116). A higher value of R will mean that it will have to be darker before the selector switch will operate.

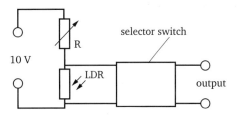

**Figure 1.116**

## Potentiometers

Connecting a **potentiometer** across a supply allows any output voltage to be selected (up to the limit of the supply being used) (Figure 1.117).

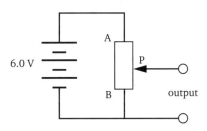

**Figure 1.117**

The potentiometer usually consists of a long resistance wire, AB, wound in a coil with a sliding contact, P, which can be moved along the wire.

The output voltage depends on the position of P:
- with P at A, the output voltage is 6.0 V
- with P moved to B, the output voltage is 0 V
- with P moved to the midpoint of AB, the output voltage is 3.0 V.

### Test your understanding

1. Draw circuits to show selector switches operated by potential dividers so that the switch is turned on when:
   (a) the light level is high
   (b) the temperature is low
   (c) the temperature is high.

continued

2

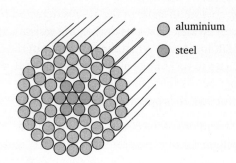

**Figure 1.118**

Figure 1.118 shows an electricity supply cable made up of seven strands of steel and 54 strands of aluminium. Each strand has a diameter of 3.0 mm.

(a) Calculate the total resistance of the aluminium strands for a 1.0 km length of electricity supply cable. The resistivity of aluminium is $2.5 \times 10^{-8}$ Ω m.

(b) The central group of steel strands has a resistance of 4.0 Ω for a 1.0 km length. By considering the steel strands and the aluminium strands to be two resistors in parallel, calculate the total resistance of 1.0 km of the cable.

(c) When the cable is in use, the resistance increases. Explain briefly, in terms of electron motion, why this happens.

3  Figure 1.119 shows a supply of emf 6.0 V and internal resistance 5.0 Ω delivering power to a pair of resistors.

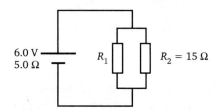

**Figure 1.119**

Maximum power is dissipated in the external resistors when the resistance of the external resistors is equal to the internal resistance of the supply.

(a) (i) Determine the value of $R_1$ that results in the maximum power being delivered to the external circuit.
(ii) Calculate the current in the circuit when the supply is delivering maximum power to the external circuit.
(iii) Calculate the potential differences across the internal resistance and across the supply terminals for the value of current you have calculated.
(iv) Calculate the power dissipated in the 15 Ω resistor when the supply is delivering maximum power to the external circuit.

(b) The 15 Ω resistor is changed for one with a resistance of 4.0 Ω. Explain why the changed circuit cannot deliver maximum power to the external circuit, whatever the value of $R_1$.

(c) (i) The 15 Ω resistor is made from wire of length 2.3 m. The wire has a diameter of $3.0 \times 10^{-4}$ m. Calculate the resistivity of the material from which it is made.
(ii) The same material is used to make wires of the same length with diameters of up to $10 \times 10^{-4}$ m. Draw a graph of the variation of resistance with diameter for the wires.

4  The resistance of a 12 V lamp filament is 6.0 Ω when it is operating at normal power. Calculate:
(a) the current in the lamp
(b) the power produced by the lamp
(c) the drift velocity of the electrons through the filament. (The filament has a diameter of $4.0 \times 10^{-5}$ m, the number of electrons per m³ is $2.0 \times 10^{28}$ and the charge on an electron is $1.6 \times 10^{-19}$ C.)
(d) the resistivity of the material from which the filament is made. (The length of the filament is 0.18 m.)

5

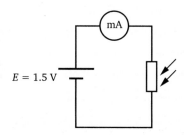

**Figure 1.120**

The current through an LDR is monitored. Figure 1.121 shows how the current varies over a period of 10 minutes.

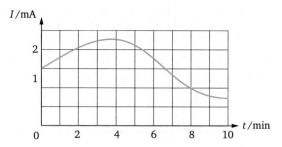

**Figure 1.121**

(a) Calculate the minimum resistance of the LDR.
(b) Calculate the amount of charge drawn from the cell during the period covered by the graph in Figure 1.121.
(c) Explain why the current through the LDR might vary in the way shown by Figure 1.121.
(d) Calculate the energy supplied by the cell.

# Unit 2 Waves

## What is a Wave?

Mechanical energy can be transferred from one place to another in two ways:
- A particle can be given kinetic energy in one position and gives up its energy when it arrives at another (e.g. throwing a ball as shown in Figure 2.1).

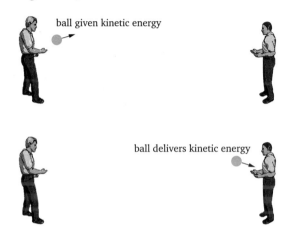

**Figure 2.1**

- A particle or object A is given energy that it passes on to another nearby particle or object B (e.g. by means of a collision). B passes the energy on to C, etc. The result is what occurs in wave motion.

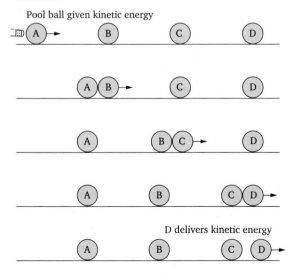

**Figure 2.2**

As shown in Figure 2.2, the particles themselves are displaced only slightly from their original position. The particle that delivers the energy is not the same one that was given the energy in the first place.

## Making Waves

All waves originate with a disturbance. In mechanical waves the disturbance is a movement of the source.

A disturbance such as a sudden single pressure change (bursting a balloon) produces a wave in the form of a pulse.

An object that is oscillating (or vibrating) periodically produces a continuous wave.

## Periodic Motion

This is motion in which the variation of displacement with time repeats itself. Figure 2.3 shows a graph for such a motion.

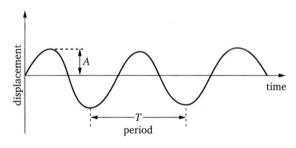

**Figure 2.3**

Practical examples of periodic motion are vibrations of the prongs of tuning forks, strings in musical instruments and air in pipes.

(Note that periodic motion is investigated experimentally in Unit 1 and analysed in detail in Unit 4.)

## Definitions

**Frequency** ($f$) is the number of complete oscillations of a particle each second. It is measured in Hz:

$$1 \text{ Hz} = 1 \text{ complete cycle per second}$$

**Period** ($T$) is the time taken for one complete oscillation.

**Amplitude** ($A$) is the maximum displacement of a particle from its equilibrium position. The greater the amplitude, the greater the energy transferred by the wave.

### Relationship between $f$ and $T$

There are $f$ oscillations per second. One oscillation takes $1/f$ seconds. This, by definition, is the period of the oscillation:

$$T = \frac{1}{f}$$

## Oscillations of Particles in Waves

When a wave that originates with an oscillation travels through a medium, each particle also oscillates with a similar motion to that of the source.

The frequency of oscillation is the same for all particles. The particles vibrate in different **phases**.

The amplitude may become smaller because of energy losses in the medium (the temperature may rise). The amplitude may become smaller because the energy spreads out over a wider area.

### Snapshot of a Wave

The black line in Figure 2.4 shows how the displacement of different parts of a medium carrying a pulse varies with distance from a source at a particular time. Figure 2.5 shows a similar snapshot for a continuous **sinusoidal** wave.

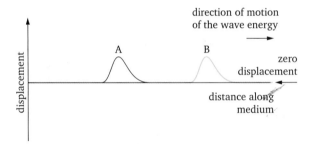

**Figure 2.4**

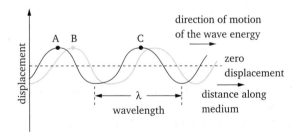

**Figure 2.5**

Note that although Figures 2.3 and 2.5 show similar shapes, the x-axis represents different quantities.

In Figure 2.5 the length $\lambda$ is called the **wavelength**.

The wavelength is the shortest distance between two parts of the wave that are oscillating in phase with each other.

### Travelling Waves

The blue lines in Figures 2.4 and 2.5 show the waves a little later. The particles with maximum displacement are now further along the medium. The energy has travelled from A to B. Since the energy is moving along the medium the wave is a **travelling wave**.

### What is a Wave Front?

A **wave front** is a way of representing a wave that is travelling in two dimensions (e.g. waves on the surface of water) or three dimensions (e.g. sound or light waves). When waves are produced by a 'point' dipper to produce waves in a ripple tank, circular waves are sent out as shown in Figure 2.6.

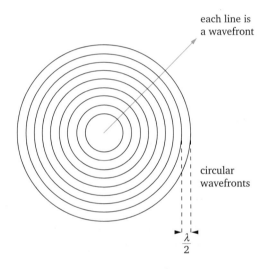

**Figure 2.6**

The circular waves seen travelling along the water are wave fronts. At the seaside a wave front is a single wave arriving at the beach.

A wave front is a line that joins all adjacent points that are at the same part of their oscillatory motion. Two-dimensional waves are usually shown as a series of lines representing wave fronts.

When wave fronts are parallel the wave is called a **plane wave**. Figure 2.7 shows a plane wave. Waves that have travelled a long way from a point source effectively become plane waves.

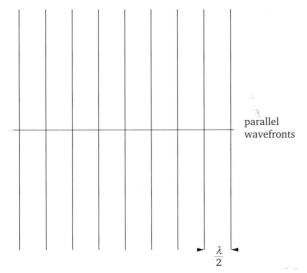

**Figure 2.7**

The way in which we normally represent a wave is shown by the lines in Figure 2.5. This is the cross-section of the wave viewed from the side along a line shown by the arrow in Figures 2.6 and 2.7.

## Relationship between Velocity, Wavelength and Frequency

The velocity $v$ of a wave is the rate at which energy is transferred in a particular direction by the wave. This is the same as the velocity at which a wave front (a peak or trough of the wave) moves through the transmitting medium.

When the maximum displacement (wave peak) A in Figure 2.5 has moved to point C the wave will look exactly the same.

During this time each particle in the medium will complete one cycle of oscillations.

The distance AC is one wavelength, $\lambda$.

The time for one oscillation is the period $T = \dfrac{1}{f}$.

The speed of the wave is:

$$v = \dfrac{\text{the distance AC}}{\text{the time for the wave to move from A to C}} = \dfrac{\lambda}{T}$$

Hence $v = f\lambda$.

## Phase Difference

Particles that reach their maximum displacement at the same time are said to be oscillating **in phase**.

When particles with the same frequency do not reach their maximum displacement at the same time they are said to be oscillating **out of phase**.

The **phase difference** between oscillating objects is referred to by the angle difference rather than a difference in terms of the time period or fractions of wavelength.

The waveform in Figure 2.8 shows two cycles of a wave. It is similar in shape to that for the sine of an angle, plotted on the $y$-axis, against the angle, plotted on the $x$-axis. (You may wish to prove this for yourself using a spreadsheet or calculator.)

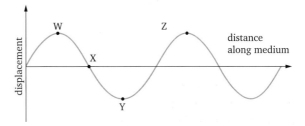

**Figure 2.8**

Particles W and X are oscillating 90° out of phase. X will reach its maximum a little later $\left(\dfrac{T}{4}\right)$ than W.

W and Y are 180° out of phase. W and Z are oscillating in phase.

Figure 2.9 shows the snapshot picture for two waves that are travelling in the same direction (a) in phase, (b) with a phase difference of 90° and (c) with a phase difference of 180° (antiphase).

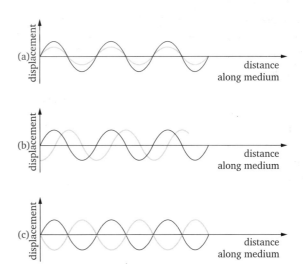

**Figure 2.9**

**Test your understanding**

1. A particle in a wave oscillates with a period of 1.5 s. With the aid of a sketch, explain what this means.
2. Calculate the periods of oscillation for the following frequencies:
   (a) 2.0 Hz  (b) 50 Hz  (c) 250 Hz
   (d) 30 kHz  (e) 25 MHz  (f) 4.0 GHz
3. Calculate the frequencies that correspond to the following periods. Give your answers to appropriate significant figures in both standard form and using appropriate prefixes.
   (a) 0.25 s  (b) 1.4 ms  (c) 120 s
   (d) 5.0 µs  (e) 48 ns  (f) 8.34 × $10^{-3}$ s

continued

4. Draw a sketch to show how the displacement of the medium varies with distance from the source for a square wave. Indicate the amplitude and wavelength on your sketch.
5. State two practical examples of periodic motion other than those mentioned on page 53.
6. Draw a labelled graph to show three cycles of a wave that is approximately sinusoidal and has an amplitude of 0.060 m and a wavelength of 0.30 m.
7. Figure 2.10 shows the displacement–time graph for a particle in a wave.

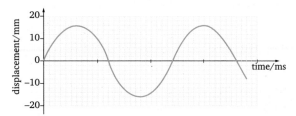

**Figure 2.10**

(a) Determine the period of oscillation of the particle.
(b) Determine the amplitude of the oscillation.
(c) Determine the frequency of the oscillation.
(d) Copy the diagram and add the displacement–time graph for a particle that is oscillating 45° out of phase with this particle.

8. Copy Table 2.1 and fill in the gaps.

**Table 2.1**

| Frequency | Velocity | Wavelength |
|---|---|---|
| 15 Hz | 120 m s$^{-1}$ | |
| | 5000 m s$^{-1}$ | 1.3 mm |
| 1.2 × 10$^{15}$ Hz | 2.0 × 10$^{8}$ m s$^{-1}$ | |
| 230 Hz | | 1.48 m |
| | 3.0 × 10$^{8}$ m s$^{-1}$ | 2.8 cm |

9. A wave travels at 330 m s$^{-1}$. The wavelength is found to be 2.4 m.
Calculate:
(a) the period of oscillation of a particle in the medium transmitting the wave
(b) the frequency of oscillation of the source that produced the wave.

## Complex Waves

Figure 2.3 represents the simplest form of periodic motion (**sinusoidal**) and this gives rise to the wave profile shown in Figure 2.5. Speech and music are produced by more complex oscillations. An example of a complex oscillation is shown in Figure 2.11.

**Figure 2.11**

A complex oscillation can be shown to result from the additions (or superposition) of sinusoidal oscillations that have different frequencies, amplitudes and phases.

Knowing what frequencies make up a complex wave is important in deciding an appropriate base bandwidth for a communication channel (see page 117).

## Transverse Waves

Energy may be transferred by wave motion as either a transverse or a longitudinal wave.

In **mechanical transverse waves** the oscillations of particles in the medium that is transferring the energy are at right angles to the direction in which the energy is being transmitted (see Figure 2.12).

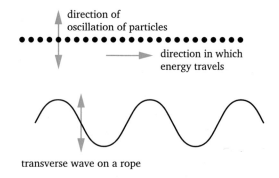

**Figure 2.12**

This type of waves is set up, for example, when the end of a long spring or rope is made to oscillate in a direction perpendicular to its length. The tension in the spring, produced as it moves upward, pulls the next part of the spring up, transferring energy to the next part of the spring and so on.

## Longitudinal Waves

In a **mechanical longitudinal wave** the oscillations of particles in the medium that is transferring the energy are in the direction in which the energy is being transmitted (see Figure 2.13).

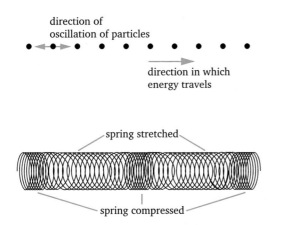

**Figure 2.13**

This type of wave is set up when the end of a long spring is oscillated in the direction of its length. When there is forward movement the force produced as the spring compresses causes the next part of the spring to move forwards, and so on. When there is backward movement the reverse occurs.

## Sound Waves

Sound in a gas, liquid or solid travels as a **longitudinal wave**.

As the sound source moves in the direction of travel of the wave energy, a **compression** is produced. This is a region where the pressure is higher than normal and is caused by atoms or molecules being closer together.

The pressure increase causes the atoms or molecules in front of the compression to move forwards so that energy is transmitted.

When the sound source moves backwards the pressure falls, producing a **rarefaction** (a low-pressure region). The atoms in front of the wave are now moved backwards.

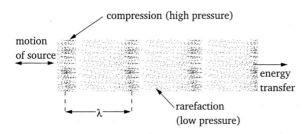

**Figure 2.14**

The net effect is that variations of pressure (the compressions and rarefactions) travel along the medium and energy is transferred from one part of the medium to the next.

## What are Light Waves and Radio Waves?

Light and radio waves are **transverse waves**. They are **electromagnetic waves** (see page 82). Unlike mechanical waves they do not require a medium to transfer the energy and can travel through an evacuated space at $3.0 \times 10^8$ m s$^{-1}$.

Radio waves originate by electrons oscillating in an aerial. This produces a varying electric field that is transmitted to the next part of space.

Light waves originate from transitions of electrons in atoms (studied in A2).

## Wave Speeds

The **wave speed** is the distance travelled by the energy in 1 s. This is the distance travelled by a particular peak or trough of the wave. Knowing wave speeds is important in many applications.

## Radar

The transmitter emits short pulses of microwave radiation. The time taken for a pulse to travel from the transmitter to an object and back again at a speed of $3.0 \times 10^8$ m s$^{-1}$ is measured. The time is converted into a total distance travelled by the pulse using $d = vt$.

The distance of the object from the transmitter is half of this distance (since the pulse has to cover the distance between the transmitter and the object twice).

The time taken for a pulse to travel to and from a reflecting object 1 km away is 6.7 μs.

## Ultrasound

This is sound with frequencies greater than 20 kHz. It is used to detect underwater objects, faults in metals and abnormalities in the body.

The speed of ultrasound varies with the type of material through which the sound travels. It is about 1500 m s$^{-1}$ in water and 5100 m s$^{-1}$ in aluminium.

The time $t$ taken for ultrasound travelling at a speed of $v$ to travel to a reflector and back is measured. As with radar the distance to the reflector is $vt/2$.

In both ultrasound and radar the time between the transmission and reception of a pulse can be measured using an oscilloscope (Figure 2.15).

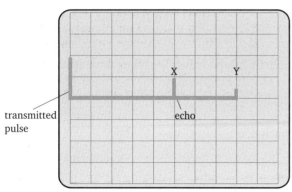

time base set to 5 μs cm$^{-1}$

**Figure 2.15**

In Figure 2.15 the time for the trace to move 1 cm across the oscilloscope screen is 5 μs (fixed by the time base of the oscilloscope). X is the reflection from an object that arrives after 25 μs. X is 25/6.7 = 3.7 km away from the transmitter.

**Test your understanding**

1. In Figure 2.15, Y is the reflection from another object.
   (a) How far away is the object?
   (b) Give two possible reasons for the smaller pulse received from Y compared with that from X.

Usually in radar the transmitter scans the area. Pulses are sent out at regular intervals as the transmitter moves through an arc or even a complete circle. An image is built up showing a two-dimensional picture of the region with all the objects (ships, aircraft or land mass) that are reflecting the pulses of energy.

Ultrasound images are built up in a similar way. The sound pulses quickly scan different parts of the body, building up a picture of an organ as it does so. Images may be single images of an organ or moving images showing, for example, the functioning of a heart.

## Measuring the Speed of Sound

Wave speeds are determined by measuring the time taken for wave energy to move through a measured distance. Pulses or continuous waves may be used. Two experiments for measuring the speed of sound in free air in a laboratory are described below. There are other methods. You need to know one.

**Experiment 1**

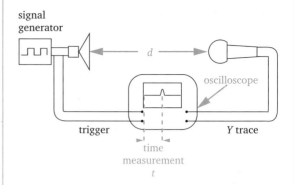

**Figure 2.16**

The loudspeaker in Figure 2.16 is driven by a low-frequency square wave from a signal generator. This produces a short pulse of waves at the natural frequency of the loudspeaker cone.

The pulse from the signal generator is used to trigger the oscilloscope. The trace commences when the square wave becomes positive.

The pulse is detected by the microphone and appears as a trace on the oscilloscope. The time taken for the pulse to travel from the speaker to the microphone is measured using the time-base scale.

The distance between the front of the speaker and a suitable point on the microphone is measured. The microphone is then moved to a new position and the new distance and corresponding time are measured. This is repeated until there are at least five sets of data for a suitable range of distances (measurements are repeated and average readings used).

A graph is drawn of distance ($y$-axis) against time ($x$-axis). The graph will be a straight line (Figure 2.17). The gradient of the line gives the speed of sound in air at the temperature of the laboratory.

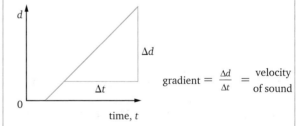

gradient = $\frac{\Delta d}{\Delta t}$ = velocity of sound

**Figure 2.17**

The graph does not go through the origin because there are systematic errors such as electrical delays in the triggering process and the unknown position of the actual source sound and points at which sound is transmitted and received.

(Any oscilloscope will work but the experiment is best conducted with a storage oscilloscope that stores and continually displays one trace. This makes the time easier to determine accurately.)

### Experiment 2

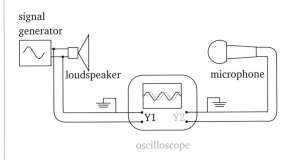

**Figure 2.18**

In Figure 2.18 the signal generator drives the loudspeaker that produces sinusoidal sound waves. The signal generator is connected to the Y1 trace of a double beam oscilloscope. The microphone is connected to the Y2 trace.

Two traces are observed. The position of the microphone is adjusted until the traces are in phase. The microphone is then moved until the traces are in phase again. The distance $d$ moved by the microphone is measured.

In going from one position where the waves are in phase to the next, the microphone has moved one wavelength, $\lambda$.

The frequency $f$ is determined by measuring the time for one oscillation $T$ using the oscilloscope $\left(f = \dfrac{1}{T}\right)$.

The speed of sound, $v = f\lambda$.

An accurate value for $\lambda$ can be determined by moving the microphone through a known number of wavelengths $n$. $\left(\lambda = \dfrac{\text{distance moved}}{n}\right)$

### Test your understanding

1. Using ultrasound equipment (called sonar) an echo is received from a shoal of fish 6.5 ms after the pulse was transmitted. The speed of sound in water is 1500 m s$^{-1}$.
   (a) Calculate the distance of the shoal from the transmitter.
   (b) Suggest why the echo pulse lasts longer than the transmitted pulse.
2. The data in Table 2.2 were obtained for Experiment 1.

**Table 2.2**

| Distance/m | 0.200 | 0.320 | 0.400 | 0.520 | 0.600 |
|---|---|---|---|---|---|
| Time/ms | 0.70 | 1.02 | 1.24 | 1.58 | 1.87 |

Plot a suitable graph to determine:
   (a) the speed of sound in air
   (b) the time delay between transmitting the pulse and receiving it when the measured distance is zero.
   State the causes of the delay in (b).
3. While doing Experiment 1 using a frequency of 1200 Hz a student determines a point where the waves are in phase. The speed of sound is 340 m s$^{-1}$. How far would the student have to move the microphone before the waves are again in phase?

## Properties of Waves

### Intensity and the inverse square law

For sound and electromagnetic waves (e.g. light waves, radio waves, X-rays and gamma rays) the **intensity** at a point is the energy arriving per second per square metre. It is measured in W m$^{-2}$.

The intensity $I$ depends on the amplitude $A$ of the wave at the point

$$I \propto A^2$$

An important case is when the energy comes from a **point source** (i.e. one that is small compared with the distance at which the intensity is being measured). In this case the intensity varies according to an inverse square law.

Figure 2.19 shows that as the wave energy moves away from the source S, it spreads out equally in all directions over the surface area of a sphere.

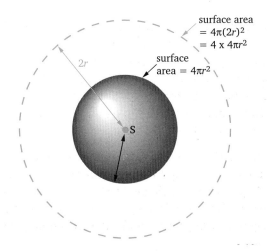

**Figure 2.19**

At a distance $r$ from the source the energy spreads over an area $4\pi r^2$.

The intensity at this distance is:

$$I = \dfrac{P}{4\pi r^2}$$

where $P$ is the power produced by the source.

The further the point is from the source, the less energy falls on a square metre. Doubling the distance results in the intensity falling by a factor of 4. Gamma radiation follows this law; this explains why keeping as large a distance as possible between the experimenter and the source reduces the radiation risk of gamma rays.

EXAMPLE: The intensity of the radiation reaching the outer atmosphere of the Earth is 1400 W m$^{-2}$. Calculate:
(a) the total power emitted by the Sun
(b) the intensity on Pluto.

Distance from Earth to Sun = $1.5 \times 10^{11}$ m
Distance from Pluto to Sun = $5.9 \times 10^{12}$ m

At $1.5 \times 10^{11}$ m the Sun's radiation is spread over a sphere of radius:

$$4\pi(1.5 \times 10^{11})^2 = 2.8 \times 10^{23} \text{ m}^2$$

Total power emitted by the Sun
$= 1400 \times 2.8 \times 10^{23}$ W
$= 4.0 \times 10^{26}$ W

Intensity on Pluto
$= 4.0 \times 10^{26}/4\pi(5.9 \times 10^{12})^2$
$= 0.90$ W m$^{-2}$

**Note**: When a wave is produced by a 'point' disturbance in water the energy spreads out over the circumference of a circle ($2\pi r$) as it moves away from the source. In this case the intensity is proportional to $\frac{1}{r}$.

## Reflection

Waves that are incident on solid surfaces obey the law of reflection. The angle between the direction of travel of the incident wave energy and the normal to the surface is equal to the angle between the direction of travel of the reflected energy and the normal (Figure 2.20).

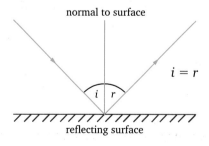

**Figure 2.20**

When reflection occurs the reflected wave appears to come from a point as far behind the reflecting surface as the object is in front of it:

- sound is reflected from large solid surfaces such as buildings and cliff faces – this produces echoes
- light is reflected from mirrors and other shiny surfaces
- radio waves are reflected by solid surfaces such as hills and large buildings – metals are particularly good reflectors.

Reflected waves travelling by different routes may interfere with each other or with waves received directly. The waves superpose and produce interference (see page 68).

## Refraction

This is the change in the direction of travel that occurs at the interface between two regions in which the wave speed is different.

When the wave speed decreases, the direction of travel moves toward the normal to the surface (Figure 2.21).

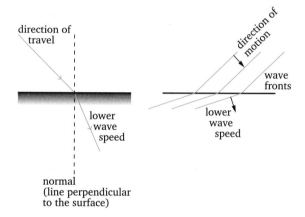

**Figure 2.21**

When the wave speed increases, the direction of travel moves away from the normal (Figure 2.22).

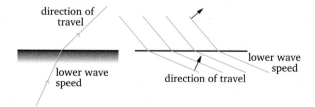

**Figure 2.22**

The degree to which the wave deviates at the interface depends on:
- the wavelength of the incident wave
- the change in wave speed that occurs.

Short wavelengths deviate most. The greater the change in speed, the greater the deviation.

## Partial reflection

When a wave is incident on an interface between two surfaces some energy is usually refracted and some is reflected. In the case of light incident on a glass surface, about 10% of the energy might be reflected and 90% transmitted.

This is why you can see your reflection in a window from indoors at night but not during the day. In the daytime the transmitted light from outside has a greater intensity than the reflected light from inside the room.

### Partial reflections when using ultrasound

In medical diagnosis it is necessary to ensure that as much ultrasound energy as possible enters the patient.

A lot of energy is reflected if the transmitter is held away from the body, as shown in Figure 2.23.

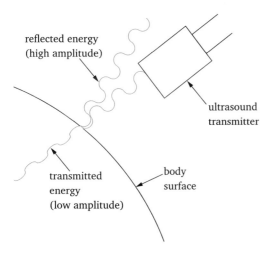

**Figure 2.23**

Even when it touches the body surface too much energy is reflected from the surface to enable good images to be formed. To increase the energy transmitted a gel is used at the interface between the transmitter and the patient, as shown in Figure 2.24.

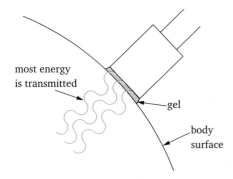

**Figure 2.24**

## Critical angle and total internal reflection

When waves travel into a medium in which they travel faster, the direction of travel deviates away from the normal.

When the incident angle becomes equal to the critical angle the energy travels along the interface.

For greater angles of incidence there is no refraction and total internal reflection occurs.

Each of these situations is shown in Figure 2.25.

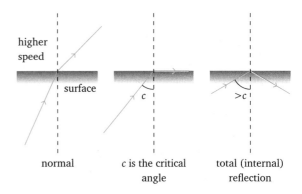

**Figure 2.25**

## Refraction in Communications

### Using the ionosphere

High-frequency radio waves may be refracted by the ionosphere. Waves that are leaving the Earth are refracted back again. This allows radio signals to be received even if the receiver is out of sight of the transmitter.

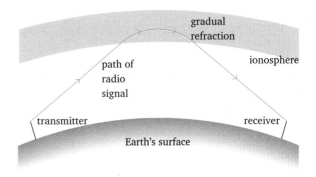

**Figure 2.26**

## Absorption

Waves lose energy as they travel through a material medium because of the absorption of energy by the medium. The energy may be lost in a number of ways depending on the type of wave used:
- sound waves cause the molecules of the medium to vibrate and energy is lost due to heating of the material

- radio waves may be absorbed by the molecules in a material, causing heating (e.g. microwaves in cooking)
- light waves may cause atoms to be excited and energy is then emitted in the form of different wavelength radiations
- X-rays (photons) may cause ionization of atoms and molecules in the material so that the intensity is reduced
- gamma rays behave like X-rays.

X-ray and gamma ray absorption is used in medicine where dense body material such as bone absorbs more X-ray energy than fleshy parts, resulting in useful shadow photographs. The energy absorbed can also be used effectively to destroy cancerous cells.

## Polarization

Polarization is a property of transverse waves. It is the name given to the process of confining a transverse wave to vibrations that are in one plane. The wave is then said to be **polarized**.

Longitudinal waves vibrate in a single direction (there is no plane of vibrations) so they cannot be polarized. Polarization is therefore a distinguishing feature of transverse waves.

### Polarization of radio waves

Radio waves are polarized in the direction of orientation of the aerial that transmits them. The up and down motion of the electrons causes a wave similar to that produced by oscillating the end of a spring or rope (Figure 2.27).

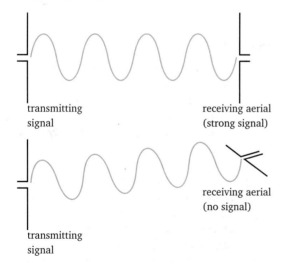

**Figure 2.27**

When a receiving aerial is oriented in the same direction as the transmitting aerial, it can pick up the transmission since electrons in the aerial are forced into oscillation by the transmitted electric field. When the receiving aerial is set up at right angles to the transmitting aerial, no signal is detected (Figure 2.27).

(Note that in practice a weak signal may be received. This is because reflected waves may change their plane of polarization.)

### Polarization of light waves

Light sources emit light that is not polarized. The vibrations are in all possible directions perpendicular to the direction of transfer of energy.

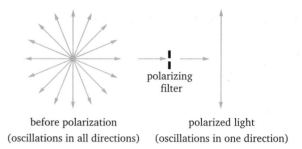

before polarization  polarized light
(oscillations in all directions)  (oscillations in one direction)

**Figure 2.28**

Figure 2.28 represents a view of what the oscillations of light waves travelling towards you from the page would be like if you could see them. Placing a sheet of Polaroid (such as is used in sunglasses) between you and the light source will only let oscillations that are in one direction pass through. The intensity of the light will be reduced.

## Fibre Optic Cables

A fibre optic cable consists of a core surrounded by a cladding in which light travels just a little faster. The critical angle is very large in this case. Light that is incident on the boundary is totally reflected back and forth, travelling down the cable as it does so. This is shown in Figure 2.29.

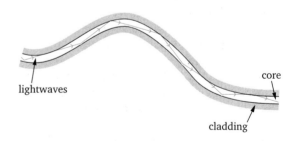

**Figure 2.29**

### Application of fibre optic cable in telecommunications

Fibre optic transmission is useful because:
- the glass can be made very pure so that there is little energy loss as the wave travels
- it is very secure as the signal can only be extracted by breaking the cable
- many more channels can be transmitted down a fibre optic cable than copper cable.

## Application of fibre optic cable in medical diagnosis

In medical diagnosis an endoscope is used to gain information about the function of internal organs without the need for surgical operations. It consists of bundles of glass fibres made into a light pipe, which can be passed down a patient's throat, for example.

A single tube contains glass fibres that:
- transmit light from the outside to illuminate the internal organs
- transmit the light reflected from internal organs back to the outside.

The light produces an image on a screen.

### Test your understanding

1. The total power transmitted by a source is 5.5 W. Assuming the source to be a point source:
   (a) calculate the intensity (power per m²) at a distance of 0.25 m from the source
   (b) determine how far away from the source the intensity will be 3 W m$^{-2}$.
2. The energy falling on the Moon is the energy on a circle of radius 1700 km.
   Use data from the example on page 60 to calculate the total energy per second falling on the surface of the Moon when it is $1.5 \times 10^{11}$ m from the Sun.
3. Suggest one circumstance in which sound energy is:
   (a) almost all reflected
   (b) almost totally absorbed
   (c) partly absorbed and partly reflected.
4. Light travelling in glass meets an interface with glass at exactly the critical angle. In which direction does the energy now travel?
5. (a) You are provided with a microwave transmitter and receiver. How would you demonstrate that the microwaves are polarized?
   (b) You are provided with similar equipment for ultrasound. Explain why you would be unable to demonstrate polarization in this case.
6. (a) Energy can be transmitted through the Earth by mechanical longitudinal or transverse waves.
   (i) Describe briefly the difference between the ways in which energy is transmitted by these waves.
   (ii) Transverse waves such as light can be polarized. Describe briefly the difference between light waves before and after polarization.
   (b) When exploring for minerals geophysicists measure the time taken for a pulse of sound, made by a small explosion, to reach a detector placed a known distance away. Figure 2.30 shows the situation when one such measurement is made. The distances are shown on the diagram.

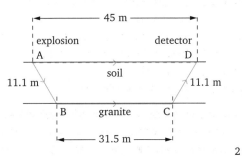

**Figure 2.30**

The speed of sound in soil is 3000 m s$^{-1}$. The speed in granite is 5000 m s$^{-1}$.
   (i) Calculate the time taken for the sound to reach the detector (a geophone) by the route AD.
   (ii) Calculate the time taken by the longer route ABCD.
   (iii) State and explain whether the detector has to be moved nearer or further from the explosion in order for the time by the short route to be equal to that by the longer route.
7. Figure 2.31 shows the profile of a wave as it moves away from a source.

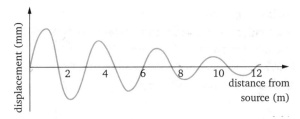

**Figure 2.31**

   (a) Determine the wavelength of the wave.
   (b) Give two reasons which might explain why the amplitude of the wave is decreasing.
   (c) The speed of the wave is 300 m s$^{-1}$. Copy Figure 2.31 and on your copy show the wave profile 3.0 ms later.
   (d) Calculate the frequency of the source.
8. (a) Describe briefly an experiment to determine the speed of sound in free air.
   (b) Explain why a health physicist includes a gel between the transmitter/receiver of ultrasound and a patient.
   (c) The time taken between the transmitted and received pulses when using ultrasound to investigate a patient is 0.33 ms. How far beneath the surface is the organ that has reflected the ultrasound?
   The speed of sound between the transmitter and the organ is 1200 m s$^{-1}$.

# Diffraction and Interference

## What is Diffraction?

The deviation or spreading of waves around an obstacle is called **diffraction** (Figure 2.32). Because of this spreading, waves do not produce perfectly sharp shadows when they meet an obstacle.

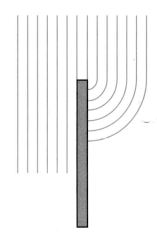

**Figure 2.32**

The sharpness of the shadow varies with:
- the wavelength of the wave
- the size of the obstacle.

Diffraction is best observed when the obstacle is of similar size to the wavelength of the wave. When light waves fall on a tree the effect of diffraction is not noticeable and a sharp optical shadow is produced. However, two people could hold a conversation round the same tree with little difficulty. Long wavelength waves diffract more.

## Waves through a gap

Figure 2.33 shows the wave energy spreading out when it passes through a gap such as a slit or hole.

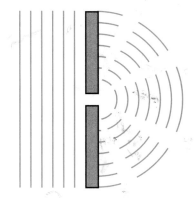

**Figure 2.33**

A graph showing how the intensity varies with the angle at which the wave is observed is shown in Figure 2.34.

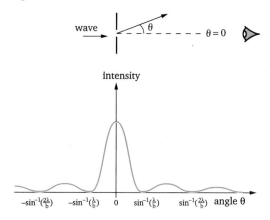

**Figure 2.34**

The intensity of the wave at a given angle is a measure of the energy per second in that direction. The light is at its greatest intensity directly in front of a slit and falls to a minimum when

$$\sin\theta = \frac{\lambda}{b}$$

where $\lambda$ = the wavelength of the wave and $b$ = the width of the slit.

This equation is also a good approximation when the wave passes through a circular hole instead of a slit. After reaching the minimum, the intensity increases again. However, most of the diffracted energy falls within the central maximum.

## Effect of slit width and wavelength

The black line in Figure 2.35 shows how intensity varies with the angle at which the pattern is viewed for a particular slit width and wavelength. The blue line shows the effect of using a narrower slit width with the same wavelength. The intensity is reduced overall because of the narrower slit and the minimum occurs at a greater angle.

Using a wave of longer wavelength with the same slit produces a broader pattern.

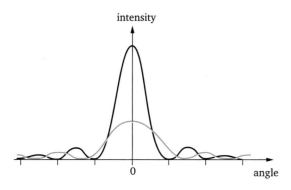

**Figure 2.35**

From Figure 2.35 it can be concluded that to diffract more energy very narrow slits are needed.

For a slit of width $b = 2\lambda$ the first minimum occurs at 45°.

For a slit of width $b = \lambda$ the first minimum occurs at 90°.

### Effects of diffraction

The following are examples of effects in which diffraction plays a role.
- Diffraction enables a radio receiver to pick up radio signals from transmitters that are not directly in its 'line of sight'.
- Diffraction influences the spreading of radio waves that are transmitted from a loudspeaker, satellite dish or radar aerial. It defines the area over which a signal is detectable.
- Diffraction is partly responsible for our limited ability to distinguish between two objects that are close together (called **resolution**).

EXAMPLE: For light of wavelength 500 nm incident on a hole of diameter $6.0 \times 10^{-5}$ m determine:
(a) the angle between the two minima on either side of the central maximum
(b) the area over which most of the energy falls at a distance of 20 cm from the hole.

(a) The minimum occurs when

$$\sin\theta = \frac{\lambda}{b} = \frac{500 \times 10^{-9}}{6.0 \times 10^{-5}} = 8.8 \times 10^{-3}$$

Angle for first minimum = 0.48°
Angle between minima = 2 × 0.48 = 0.96°

(b) Refer to Figure 2.34. The energy falls in a circle of radius $R$:

$$\frac{R}{20} = \tan 0.48° = 8.4 \times 10^{-3}$$

$$R = 0.17 \text{ cm}$$

Area on which energy falls = 0.088 cm²

## Diffraction and Radio Transmission

Figure 2.36 shows how medium and long wavelength radio waves diffract round the Earth's surface. Although the signal is weaker than a direct signal, it can be amplified using electronic circuits.

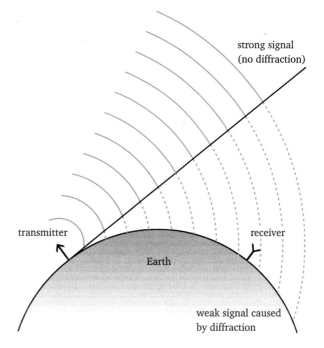

**Figure 2.36**

VHF (very high frequency) transmissions have a short wavelength, resulting in little diffracted energy. These transmissions can only be received when the receiver is in the **line of sight** of a transmitter, i.e. there is a direct route from transmitter to receiver (Figure 2.37). This explains why transmitters and receivers are positioned as high up as possible.

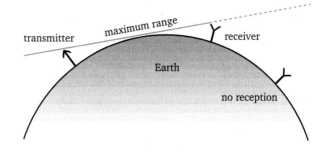

**Figure 2.37**

## Energy from Aerials

An aerial or a loudspeaker is effectively a gap through which energy radiates.

When an aerial transmits energy into a small angle, such as when transmitting signals to a satellite, a large aperture (a wide dish) is needed.

When the energy has to spread out so that more receivers can pick up the radiated energy, such as when the satellite transmits the signal back to Earth, a smaller aperture is used.

These two situations are illustrated in Figure 2.38.

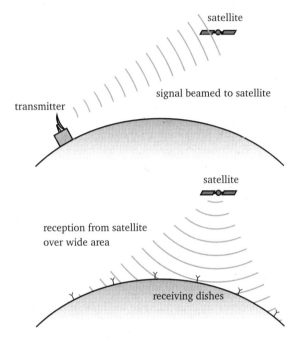

**Figure 2.38**

## Loudspeaker Design

### Where should you sit?

When sound of frequency 2400 Hz ($\lambda \approx 0.14$ m) is emitted by a loudspeaker of diameter 0.15 m, the first minimum occurs when $\sin \theta = 0.14/0.15$. The angle at which a minimum occurs is about 69°.

For a listener positioned within this angle on either side of the loudspeaker all frequencies lower than this will be heard reasonably loudly. For a simple radio this is adequate.

### Effect of different frequencies

In a high-fidelity sound system the speaker needs to emit frequencies in excess of 10 kHz ($\lambda = 0.034$ m). At 10 kHz the first minimum for a 0.15 m diameter loudspeaker occurs at 13°. Most of the energy falls within this angle. Only a listener sitting almost directly in front of the loudspeaker would hear good quality sound.

### Woofers and tweeters

To overcome the frequency problem a second loudspeaker of diameter about 0.04 m is included in a high-fidelity speaker cabinet. The lower frequencies are steered (using electronic filters) to the larger speaker (the woofer) and the higher frequencies to the smaller loudspeaker (the tweeter). The tweeter diffracts the high frequencies more because of its smaller value of $b$, giving the listener a greater choice of seating positions.

(Note that the effect of resonance is also an important aspect in loudspeaker design. This will be studied in A2.)

## Resolution

**Resolution** is our ability to differentiate between two sources of waves.

The angle subtended by two objects at the eye, when we can just tell that they are two objects, is a measure of resolution.

Light enters our eyes through a pupil of diameter about 5 mm. A typical optical wavelength (green light) is 500 nm.

The minimum angle for two objects to be resolved (seen as separate objects) occurs when the diffraction maximum of one object coincides with the diffraction minimum of the other (Figure 2.39).

The minimum angle subtended at the eye that enables two objects to be seen as separate is given by:

$$\sin \theta = \frac{\lambda}{b}$$

$$= \frac{500 \times 10^{-9}}{5 \times 10^{-3}}$$

The minimum angle for resolution is therefore $5.7 \times 10^{-3}$ degrees.

If two objects subtend an angle that is equal to or larger than this they can be resolved.

Resolution becomes worse as wavelength increases or as pupil (or aperture) size falls.

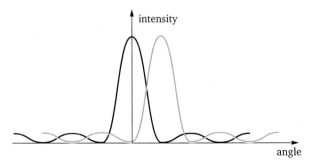

**Figure 2.39**

### How close are objects that are just resolved?

The closest objects that a normal eye can see comfortably are about 0.30 m from the eye.

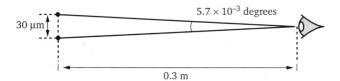

**Figure 2.40**

The two small objects in Figure 2.40, which are separated by 30 $\mu$m (0.30 sin 5.7 × 10$^{-3}$), can just be resolved at a distance of 0.3 m.

#### Test your own resolution

Draw two lines 2 mm apart on a sheet of card and determine the distance at which you can see them as two separate lines. Then work out the angle subtended at your eye by the lines.

### VDU pixels

The separation of pixels on a television screen, designed to be watched at a distance of 2.5 m, needs to be less than 0.25 mm (2.5 sin 0.7 × 10$^{-3}$) so that individual pixels are not seen by the eye.

### Telescopes

Optical telescopes have larger diameter apertures than the pupil of the eye. Resolution for optical wavelengths is improved.

Radio telescopes are designed to receive long wavelengths (e.g. 30 cm) that are emitted by astronomical objects. Even with large diameter receiving dishes (e.g. 30 m) the resolution is poor. The minimum angle for two stellar objects to be resolved is about 0.6° for the figures given above.

Two stars 100 light years away separated by one light year would be just resolved using such a telescope.

### Using ultrasound

In medical diagnosis and industrial applications efficient reflection of energy from small objects such as tumours and flaws in metals is needed. It is also necessary to have as high a resolution as possible.

Long wavelengths diffract around small objects and little energy is reflected. Most of the energy goes on as if the object were not there.

More energy and better resolution is achieved when the wavelength of the ultrasound is much smaller than the objects that have to be imaged. Frequencies of 2.5–8 MHz are used. Assuming a speed of 1500 m s$^{-1}$ this gives the wavelengths in the body as between approximately 0.2 and 0.5 mm.

### Using X-rays

X-rays have a small wavelength (10$^{-10}$ m) and therefore are diffracted only by objects the size of atoms. This enables X-rays to be used effectively to produce sharp shadows of small defects inside the body or inside metals.

Because it is diffracted by atoms, the interference of X-rays can be used to provide information about the arrangement of atoms in crystals (X-ray crystallography).

#### Test your understanding

1. Calculate the angle when the first diffraction minimum occurs when:
   (a) light of wavelength 560 nm falls on an aperture of 2 mm
   (b) light falls on a hole with a diameter of 10 times its wavelength
   (c) microwaves with a wavelength of 2.7 cm fall on a gap that is 4 cm wide.
2. Describe the effect(s) on the diffraction pattern produced by a slit when the following changes are made without making any other change:
   (a) increasing the source intensity
   (b) increasing the slit width
   (c) reducing the wavelength.
3. (a) A signal of wavelength 0.034 m is used to drive a loudspeaker of diameter 0.20 m. Determine the angle within which most of the energy of the signal will fall.
   (b) What would the answer be if the signal were used to drive a speaker of diameter 0.075 m?
4. The angular resolution of an eye of an observer is 6.0 × 10$^{-3}$ degrees. A car approaches with two headlights separated by 1.2 m. How far away will the car be when the observer is able to distinguish between the two headlights?
5. Two stars are just seen as separate by a 30 cm aperture telescope. They are 25 light years away. Assuming that the stars are viewed at a wavelength of 500 nm, determine the separation of the stars.
6. Draw a sketch to show the diffraction pattern of light from two sources that are not resolved.
7. A monitor of a computer is to be viewed at a distance of 0.5 m. The diameter of the pupil of an eye is 3.0 mm.
   (a) Determine the separation of blue pixels (wavelength = 450 nm) if they are to just merge.
   (b) Explain whether light from green pixels (wavelength = 500 nm) would be resolved.

*continued* →

8 (a) In the Hubble telescope the radiation falls on an aperture of 2.4 m. The telescope can view ultraviolet radiation of wavelength 120 nm. Calculate the resolution that is possible using the Hubble telescope.
(b) On Earth it is impossible to use ultraviolet for observations because ultraviolet radiation is absorbed by the atmosphere. Visible light has to be used. What is the resolution of a telescope of similar diameter to the Hubble telescope but designed to use visible radiation of wavelength 500 nm?

## Superposition of Waves

**Superposition** occurs in a region where two or more waves of the same type overlap.

The **principle of superposition** states that the resultant displacement at a point at a given time is the algebraic sum of the displacements of each wave.

Figure 2.41 shows two pulses travelling in opposite directions along a rope. The situation at different times is shown. When the pulses meet they superpose to produce the combined effect shown in blue.

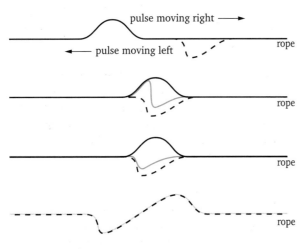

**Figure 2.41**

Note that the waves only superpose where they overlap. The waves themselves continue their passage through the medium unchanged.

## Synthesizing Complex Sounds

Any shape of waveforms can be synthesized by adding sinusoidal waveforms of appropriate frequency, amplitude and phase. This is the principle used in electronic instruments. The manufacturer determines the frequencies that make up a given sound (guitar, oboe, piano, etc.). These are generated electronically and the principle of superposition is used to add them together to generate the required sound.

## Interference

When the waves are continuous waves the displacement at the point where they overlap is continually changing. The waves are said to be **interfering**.

Under certain conditions a steady interference pattern is observed in which some points oscillate with a maximum amplitude and others either do not oscillate at all or oscillate with a minimum amplitude.

The blue line in Figure 2.42 shows how the resultant displacement at a point changes with time as two waves of different amplitudes but the **same frequency** arrive at the point **in phase**. Figure 2.43 shows what happens when the waves arrive in **antiphase**.

When waves arrive in phase, a large amplitude oscillation (the maximum possible) is produced. This is called **constructive interference**.

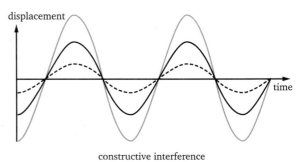

constructive interference

**Figure 2.42**

When two interfering waves have the same amplitude constructive interference produces a resultant wave of *twice* the amplitude of one of the waves.

When two waves arrive in **antiphase** the amplitude is a minimum. This is called **destructive interference**.

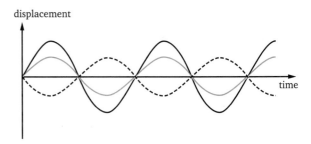

**Figure 2.43**

When waves of equal amplitude interfere destructively they cancel at all times and there is no resultant oscillation (i.e. the amplitude is zero). When they are neither in phase nor antiphase (for example, one may be 90° out of phase) there will be a resultant oscillation somewhere between the maximum and the minimum.

## Producing observable interference patterns

To produce an observable interference pattern there must always be constructive interference at some fixed points and destructive interference at others, i.e. the points at which maximum amplitude and minimum resultant amplitudes occur must always be in the same place. For this to happen the sources of the waves must be **coherent**.

The waves produced by **coherent sources** have:
- the same frequency
- constant phase difference.

Whether the coherent sources produce waves that are in phase, antiphase or have some other phase relationship does not matter as long as this does not change.

### Coherent sources

**Light** is a mixture of waves each lasting a short time and emitted by different atoms. The phase is changing all the time. Light from two separate light bulbs is therefore not coherent. Interference cannot be observed using two separate light bulbs.

**Coherent light sources** can be produced using the arrangement shown in Figure 2.44.

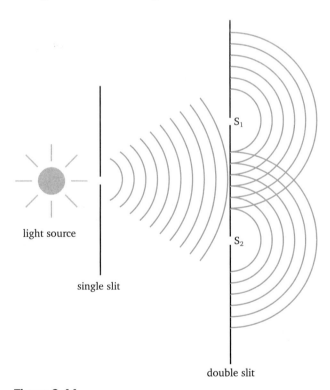

**Figure 2.44**

The light diffracts at the single slit and illuminates the double slits. Each slit diffracts the light again to produce the two overlapping beams that interfere. Because the light from each slit always comes from the same changes that takes place in atoms, the sources are coherent. This is a practical arrangement for conducting Young's two-slit experiment (not to scale).

**Coherent sound and radio sources** can be produced using the arrangements shown in Figure 2.45.

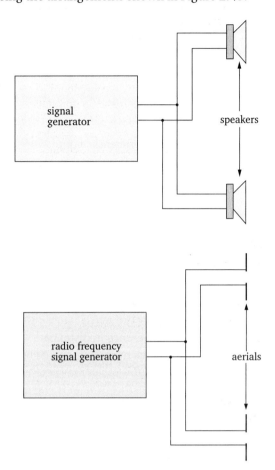

**Figure 2.45**

The waves emitted will be either in phase or antiphase, depending on how the connections are made to the loudspeakers or aerials. The waves are coherent because the loudspeaker or aerial is powered by the same electronic source.

### Test your understanding

1. Figure 2.46 shows three waveforms that are available in a simple synthesizer.

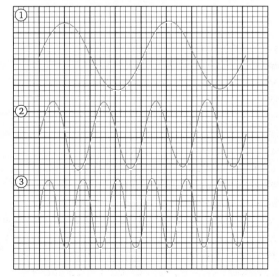

**Figure 2.46**

continued

Make appropriate copies of them and show the result of adding:
(a) the first two
(b) the first and the third

2 Draw a sketch to show the result of adding two sinusoidal waves that have the same wavelength and amplitude but are out of phase by 90°.

3 Two waves of amplitudes $a$ and $3a$ interfere. Calculate the ratio:

$$\frac{\text{amplitude at an interference maximum}}{\text{amplitude at an interference minimum}}$$

## Path Difference

To determine whether interference at a point is constructive or destructive it is necessary to determine whether waves arrive at a point in phase or antiphase. The first step is to determine the **path difference**.

Path difference is the difference in distance travelled by the two waves from their source to the point in question. The routes taken may be:
- direct, in which case the direction of travel of the wave energy is in straight lines
- indirect, in which case the wave energy may be reflected or travel along curved paths such as along cables.

In each case it is necessary only to measure or calculate the distances involved and determine the difference.

The next step is to determine how many wavelengths fit into this path difference.

If there is a **whole number** of wavelengths the waves arrive in phase and constructive interference occurs.

If there is a **whole number of wavelengths plus a half wavelength** the waves will arrive in antiphase and the interference will be destructive.

## Two-source Interference

Figure 2.47 shows wave fronts emitted from two small sources $S_1$ and $S_2$. The dotted lines represent peaks and the continuous lines represent troughs.

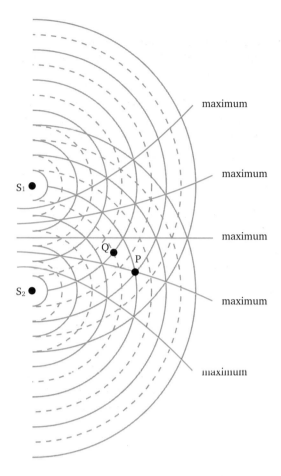

**Figure 2.47**

At the point where dotted lines or continuous lines meet there is maximum amplitude. Where a dotted line meets a continuous line there is a minimum amplitude.

The blue lines in Figure 2.47 join the points where maximum amplitude waves occur.

At point P:
- the path difference ($S_1P - S_2P$) is one wavelength
- the waves arrive in phase
- the waves interfere constructively
- there is a maximum amplitude.

Constructive interference occurs whenever the path difference is a whole number of wavelengths so

$$S_1P - S_2P = n\lambda$$

At point Q:
- the path difference ($S_1Q - S_2Q$) is half a wavelength
- the waves arrive in antiphase
- the waves interfere destructively
- there is a minimum amplitude.

Destructive interference occurs whenever the path difference is a whole number of wavelengths plus a half wavelength (i.e. an odd number of half wavelengths) so:

$$S_1Q - S_2Q = (n + \tfrac{1}{2})\lambda$$

## Monochromatic Light

Interference is easiest to observe using monochromatic light. This is light that consists of a single wavelength.

If light consisting of two or more wavelengths is used then the interference becomes confused since the positions of maximum intensity for different wavelengths occur in different places. White light (containing all visible wavelengths) is particularly poor.

Laser light is generally preferred nowadays but any light that has a dominant single wavelength, such as that produced by the discharge of electricity through sodium vapour, will produce observable patterns.

## Optical Interference: Young's Experiment

Thomas Young was the person first to set up the arrangement shown in Figure 2.48 to demonstrate interference between two coherent light sources.

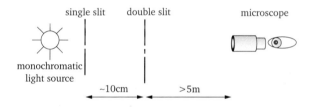

**Figure 2.48**

When monochromatic light is used, the pattern observed on a screen looks like that in Figure 2.49. The pattern of maximum and minimum brightness is referred to as **interference fringes**. The separation is the **fringe spacing**.

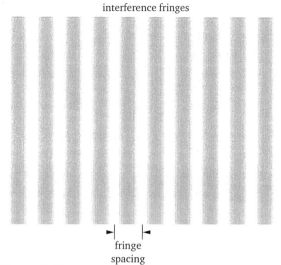

**Figure 2.49**

The fringes are faint and not easy to see without the aid of a microscope when ordinary light sources are used. A laser produces brighter fringes. Since it produces coherent light, the laser light can be used to illuminate the two source slits directly.

The fringe spacing depends on:
- the wavelength of the waves, $\lambda$
- the distance between the sources and the observation screen, $D$
- the separation of the sources, $d$.

For any type of wave, provided that $D \gg d$:

$$\text{fringe spacing, } y = \frac{\lambda D}{d}$$

### Derivation of $y = \frac{\lambda D}{d}$

Note that you will not be required to reproduce this but you should understand the principles involved as you may be asked to explain aspects of the derivation.

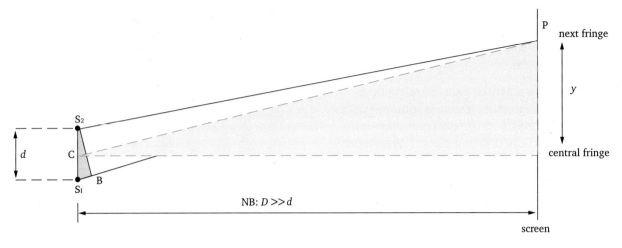

**Figure 2.50**

In Figure 2.50 the paths $S_1P$ and $S_2P$ are approximately parallel since $D \gg d$. The shaded triangles are similar (the angles are the same so the side lengths in one are proportional to the corresponding side lengths in the other).

Path difference to point $P = S_1B$.

Using similar triangles:

$$\frac{S_1B}{d} = \frac{y}{CP} = \frac{y}{D}$$

$$S_1B = \frac{yd}{D}$$

since $CP \approx D$ when the angle is small.

When this path difference is equal to $\lambda$, $y$ is the fringe spacing, therefore:

$$y = \frac{\lambda D}{d}$$

## Interference using Reflections

### Using radio waves

Interference can be conveniently investigated in a laboratory using microwaves of wavelength about 30 mm or ultra high frequency waves of wavelength about 0.3 m.

Figure 2.51 shows an arrangement using microwaves.

Some microwaves pass through the hardboard and some are reflected (i.e. there is partial reflection). All the energy incident on the metal is reflected. The waves partially reflected from the hardboard surface interfere with those that are reflected from the metal surface.

The two waves are coherent since they come from the same source. The path difference is shown by the blue lines.

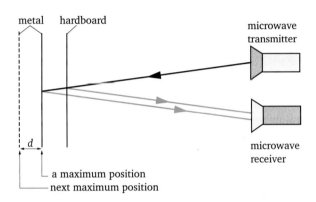

**Figure 2.51**

Suppose the metal reflector is initially positioned so that a maximum signal is observed. When the screen is moved back a distance $d$ the next maximum is observed.

The **extra path difference** is $2d$ since the wave has to travel to the metal and back again

$$\text{so} \quad 2d = \lambda$$

Every time the metal plate moves a distance $d$ the signal changes from one maximum to the next. In this way an accurate value for the wavelength can be measured.

The speed at which the metal plate is moved could be determined by measuring the frequency at which maximums are observed. This principle can be used in speed guns for measuring the speed of vehicles.

### Using light

Interference can be observed using monochromatic light incident on a thin layer of air trapped between two microscope slides, as shown in Figure 2.52.

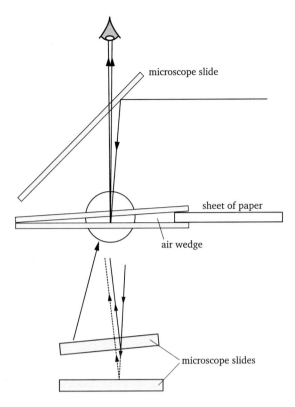

**Figure 2.52**

Waves reflected from the top and bottom interfaces of the air wedge interfere:

- a $\frac{\lambda}{2}$ change in thickness of the air wedge results in the pattern changing from one bright fringe to the next
- a change of $\frac{\lambda}{4}$ results in a path difference change of $\frac{\lambda}{2}$ so that the pattern changes from a bright to a dark fringe.

If the wedge is illuminated with white light the different colours (different wavelengths) produce maximums at different angles. This results in spectra being observed when looking at thin films of liquid (e.g. films of oil on water or films of washing up liquid) in daylight.

## Interference between Waves in Communication

Interference can occur between a signal that travels directly to a receiving aerial and one which is reflected off a large building or a hill (Figure 2.53).

The signal will be weak if the path difference is $(n + \frac{1}{2})\lambda$.

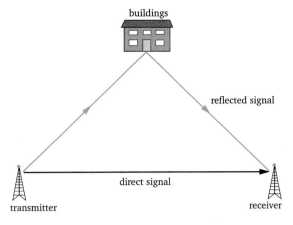

**Figure 2.53**

Reception by a car radio may vary as the car moves through positions of maximum and minimum signals.

Signals may be received directly and by refraction at the ionosphere as shown in Figure 2.54.

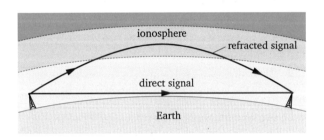

**Figure 2.54**

The signal may be strong or weak depending on the position of the receiver.

At times the ionosphere moves up and down. This causes the resultant signal strength to fluctuate at the same frequency as the ionosphere moves.

Aerials use interference to improve signal strength and to make the aerial directional.

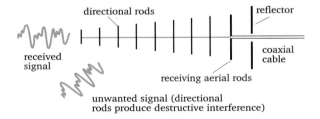

**Figure 2.55**

In the television aerial shown in Figure 2.55 the short metal bars at the front produce waves that interfere destructively at the aerial for signals coming from the side. The reflector at the back produces constructive interference for the signal coming from the direction in which the aerial is pointing.

**Test your understanding**

1. Two coherent sources A and B emit light waves of wavelength 600 nm that are in phase. Using the information below, determine whether a maximum or minimum intensity will be observed at a point P in the following instances. Show your working clearly.
   (a) The path difference between waves from A and B to P is 2.40 $\mu$m.
   (b) A is 1.5 $\mu$m further from P than B.
2. Monochromatic light falls on two slits that are 0.035 mm apart. The separation of the fringes is 8.5 $\mu$m when the distance between the slits and a screen is 0.45 m. Calculate the wavelength of the light used.
3. The screen on which the fringes in question 2 are observed is moved back to 0.90 m. How far apart are the fringes now?
4. (a) What would be the angular separation of adjacent fringes when light of wavelength 650 nm is incident on two slits that are 0.42 mm apart?
   (b) How far from the slits would a screen have to be placed to produce a fringe separation of 2.0 mm?
5. Microwaves are used to investigate interference using the arrangement in Figure 2.51. A position where a maximum signal is observed is found. The screen is then moved towards the hardboard. When moved 6.4 cm, four further maxima are observed, the final position being a maximum. Calculate the wavelength of the microwaves.
6. Using the arrangement in question 5, how far would the metal reflector have to be moved for the signal received to change from a maximum to a minimum signal?
7. Using an air wedge a dark fringe is observed using light of wavelength 590 nm. By how much must the separation of the wedge increase for the next dark fringe to be observed?
8. Coherent light of wavelength $4.80 \times 10^{-7}$ m emerges from two slits X and Y to produce an interference fringe pattern on a screen as shown in Figure 2.56.

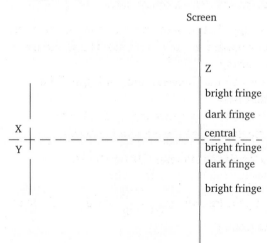

**Figure 2.56**

A dark fringe occurs at Z. Determine the path difference between the light from the two slits arriving at Z.

9. Two radio transmitters emit waves of wavelength 300 m in phase. A receiver an equal distance from each of them receives a strong signal. The receiver then moves along the line joining the two transmitters.
   (a) What will be the path difference between the waves from the two transmitters when a maximum is next received?
   (b) How far will the receiver have to move for a minimum signal to be received?
10. Figure 2.57 shows an arrangement for investigating interference using microwaves. The two vertical slits $S_1$ and $S_2$ are at equal distances from the transmitter. A receiver is moved along the line AB. Figure 2.58 shows how the amplitude of the received signal varies with position along AB.

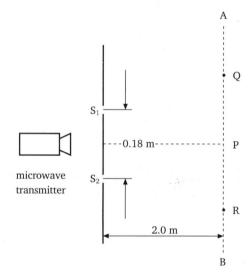

**Figure 2.57**

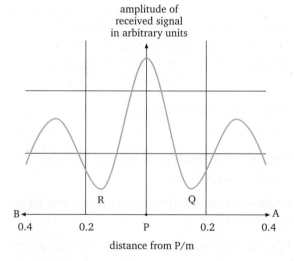

**Figure 2.58**

(a) Explain why there is:
   (i) a maximum signal at P
   (ii) a minimum signal at Q.
(b) The centres of the slits are 0.18 m apart and P is 2.0 m from the slits. The speed of microwaves is $3.0 \times 10^8$ m s$^{-1}$. Determine:
   (i) the wavelength of the microwaves
   (ii) the frequency of the microwaves.

# Diffraction Gratings

Diffraction gratings produce interference in a similar way to two slits and are used to study optical spectra. Applications include obtaining information about stars and chemical analysis.

Because of the way diffraction gratings are made the slit separation $d$ can be made small. This produces widely spaced positions of maximum intensity. The positions of maximum light intensity depend on:
- the wavelength, $\lambda$
- the slit separation, $d$.

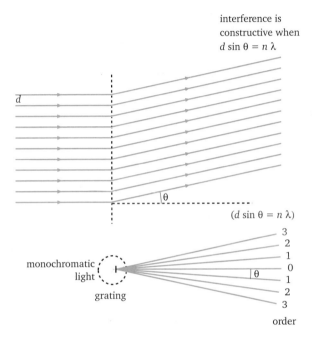

**Figure 2.59**

Figure 2.59 shows light that is incident normally (i.e. perpendicularly) on a grating. The rays can be focused to a point using a lens. The angles $\theta$ at which maximum light intensity is observed are given by:

$$d \sin \theta = n \lambda$$

where $n$ is a whole number.

The number $n$ is called the order of the maximum.

When $n = 0$ the maximum is at an angle of 0°. This is called the zeroth order.

When $n = 1$ the maximum is called the first-order maximum and so on.

The first order maximum occurs at an angle given by

$$\sin \theta = \frac{\lambda}{d}.$$

The second order maximum occurs at an angle given by

$$\sin \theta = \frac{2\lambda}{d}.$$

There are first-order, second-order, etc. maxima on both sides of the zeroth order.

## Advantages of using a diffraction grating

When compared with two-slit interference a diffraction grating produces interference maxima that are:
- brighter
- sharper.

They are **brighter** because there are more slits to let through the light. They are **sharper** because light waves only interfere constructively within a narrow angle. The more slits there are to interfere, the sharper the fringes.

### Converting lines per mm to slit spacing, $d$

A diffraction grating consists of many slits that are close together. The specification of a particular grating is the number of slits per mm. This is usually referred to as the number of lines per mm.

If there are $N$ lines per mm the distance $d$ between the slits in the grating is $\frac{1}{N}$ mm or $\frac{1 \times 10^{-3}}{N}$ m.

This is the number you insert in the grating formula.

### Using gratings practically

The practical use of gratings is to measure the wavelength of light from a source.

The angles at which maximum intensity occur depend on the wavelength. When a source emits different wavelength light a spectrum is produced. The longer the wavelength, the greater the angle at which a maximum occurs in a given order. Red light produces positions of maximum intensity that are more widely spaced than blue light.

### Measuring the slit spacing

Although gratings are sold with the number of lines per mm stated, this can be checked using light of known wavelength such as that emitted by a laser. The spectrum from a laser can be projected onto a screen as shown in Figure 2.60.

The angle between the straight-through position and a maximum (of known order) is found by measuring:
- the distance between the grating and the screen $D$
- the distance $y$ between the zeroth order maximum and the maximum chosen.

The angle is then found by $\tan \theta = \frac{y}{D}$.

Using the formula $\sin \theta = \frac{2\lambda}{d}$ a value of $d$ can be determined.

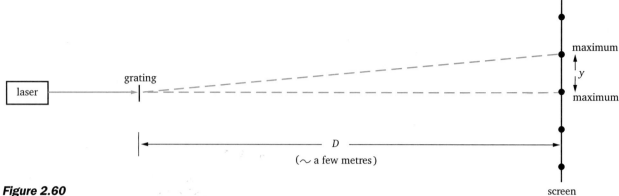

**Figure 2.60**

### Measuring unknown wavelengths

When the slit spacing has been determined the experiment described above can be repeated to determine the wavelengths of light emitted by other bodies.

### Test your understanding

1. A diffraction grating is marked with 450 lines per mm. Calculate the separation of the slits on the grating in m.
2. A diffraction grating is found to have a slit separation of $1.70 \times 10^{-6}$ m. How should it be marked in lines per mm?
3. Monochromatic light of wavelength $5.0 \times 10^{-7}$ m is incident perpendicularly on a diffraction grating. The second order diffraction lines are formed at angles of 30° from the perpendicular. Calculate:
   (a) the separation of the lines on the grating
   (b) the number of lines per mm for the grating.
4. Calculate the angles at which a maximum will be observed when light of wavelength $6.5 \times 10^{-7}$ m is incident perpendicularly on a grating that has a line separation of $2.5 \times 10^{-6}$ m.
5. Calculate the highest order maximum that is visible using a grating of spacing $5.0 \times 10^{-6}$ m when using light of wavelength $6.2 \times 10^{-7}$ m.
6. Light consisting of two wavelengths, 560 nm and 580 nm, falls perpendicularly onto a grating that has 300 lines per mm.
   (a) Calculate the spacing of the slits on the grating.
   (b) Calculate the angular separation of the maximum signal produced by each wavelength in the first order.
   (c) How far apart will these maxima be when they fall on a screen 1.5 m from the grating?
7. (a) State the conditions necessary for overlapping beams of light from two sources to exhibit interference.
   (b) Figure 2.61 shows light hitting the face of a glass lens at approximately normal to the surface.

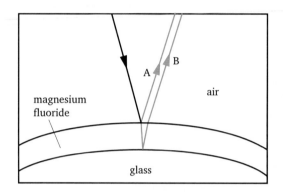

**Figure 2.61**

The lens is coated with a thin layer of magnesium fluoride to reduce the amount of light reflected from the surface. Light of frequency $5.3 \times 10^{14}$ m, reflected from the front of the layer, interferes destructively with that from the back surface. The speed of light in magnesium fluoride is $2.1 \times 10^{8}$ m s$^{-1}$. Assume that the light does not change phase when it is reflected.
   (i) Calculate the wavelength of the light in magnesium fluoride.
   (ii) Suggest a value for the thickness of the magnesium fluoride layer required for destructive interference to occur. Explain your answer.
8. (a) Interference patterns can be observed when light waves from two coherent sources overlap. The pattern is easier to observe when the amplitudes of the waves from the two sources are similar.
   (i) Explain briefly what is meant by coherent sources.
   (ii) Why is the pattern easier to observe when the amplitudes are similar?
   (b) Figure 2.62 shows an arrangement that can be used to observe interference patterns.

continued ◗

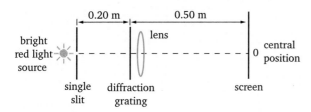

**Figure 2.62**

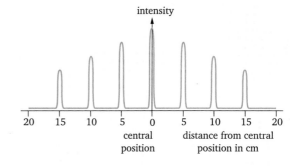

**Figure 2.63**

Figure 2.63 shows how the light intensity varies with position on the screen.

(i) The monochromatic red light used has a wavelength of 656 nm. Determine the number of lines per metre on the diffraction grating.

(ii) The monochromatic source is replaced by a source that emits radiation of wavelength 656 nm together with light of a shorter wavelength. Make a sketch to show how the intensity varies with position now. Show clearly which parts of the diagram show red intensities and which parts blue.

# Stationary Waves

## Production of a Stationary Wave

A stationary wave is produced by the superposition of two waves that:
- are the same type
- have the same frequency
- travel in opposite directions in the same medium.

The waves will have the same velocity and the same wavelength. In most practical cases the waves will also have the same or very similar amplitude, so antiphase waves will cancel completely.

Figure 2.64 shows the sequence of events at time intervals that occur when two transverse waves travel in opposite directions.

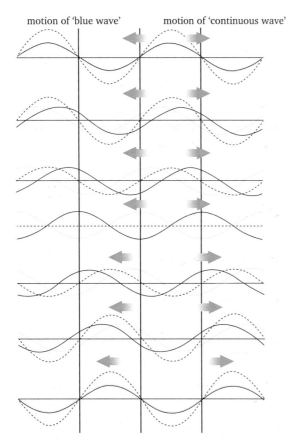

**Figure 2.64**

The diagrams show the position of each wave and the resultant wave at intervals of $\frac{1}{12}$ of the period $T$ of oscillation of a particle in the medium. The black continuous line represents the wave moving to the right and the blue line represents the wave moving to the left. The dotted line is the resultant wave at that time.

In each time interval $\frac{T}{12}$ one wave moves $\frac{\lambda}{12}$ to the right and the other $\frac{\lambda}{12}$ to the left.

## Nodes and Antinodes

The dotted line in Figure 2.64 shows that:
- at some points the displacement is always zero (the amplitude is zero)
- at other points there is a maximum amplitude.

The points that have zero amplitude are called **nodes**.

The points that have maximum amplitude are called antinodes.

The diagrams in Figure 2.64 show that the distance between two adjacent nodes or antinodes is $\frac{\lambda}{2}$.

## How do the Different Parts of the Medium Move?

Figure 2.64 shows that the oscillations of all parts of the medium between two nodes oscillate in phase but have different amplitudes of oscillation.

The oscillations of parts of the medium on one side of a node are antiphase to those on the other.

## Stationary Waves on a Stretched String

A stretched string is one that is under a tension and is arranged such that both ends are fixed. The fixed ends must be nodes because they cannot move.

Stationary waves are formed by interference between waves travelling toward the fixed ends and waves reflected from the fixed ends.

The only possible stationary waves are those for which a whole number of half wavelengths fit exactly between the fixed ends. Each possible stationary wave is called a **mode of vibration**.

The first three possible modes of vibration are shown in Figure 2.65.

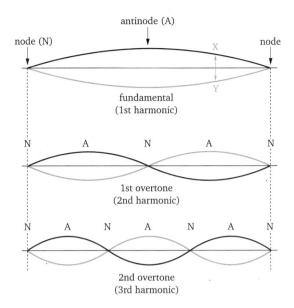

**Figure 2.65**

## Fundamental Frequency of a Stretched String

The simplest mode of vibration shows one loop between the fixed ends. Since this is the simplest mode of vibration it is called the **fundamental mode**. It is also referred to as the **first harmonic vibration**.

The wavelength of the fundamental is twice the distance between the fixed ends:

$$\lambda = 2l$$

Along a stretched string the velocity is given by:

$$v = \sqrt{\frac{T}{\mu}}$$

where $T$ is the tension in the string and $\mu$ is the mass of one metre of the string.

The frequency of the fundamental vibration is therefore given by: $v = f\lambda$

$$f = \frac{1}{2l}\sqrt{\frac{T}{\mu}}$$

You do not need to remember this formula. It will be given in questions when required.

## Overtones or Higher Harmonics

The diagram in Figure 2.65, which shows two loops between the fixed ends, represents the **first overtone** or **second harmonic vibration**.

The wavelength of this vibration is $l$ so the frequency is $\frac{1}{l}\sqrt{\frac{T}{\mu}}$. This is twice the frequency of the fundamental.

The diagram in Figure 2.65 with three loops represents the second overtone or third harmonic.

Its wavelength is $\frac{2}{3}l$ and its frequency is $\frac{3}{2l}\sqrt{\frac{T}{\mu}}$. This is three times the fundamental frequency.

Equations for the frequencies of higher harmonics can be determined in a similar way.

## Frequencies of Modes of Vibration

A stretched string can vibrate with frequencies $f$, $2f$, $3f$, $4f$,..., $nf$. When a string is plucked or struck the fundamental is the predominant frequency but other frequencies may also be present. The harmonics present when a string is bowed are different from those when it is plucked.

If the true tonal quality of the plucked or struck string is to be transmitted by radio or using a compact disc all the harmonics within the range of human hearing need to be transmitted (see bandwidth).

## How does a Stretched String Vibrate?

If the string is plucked cleanly in one plane then it vibrates in that plane. Figure 2.66 shows the vibrations of the third harmonic.

In practice the vibration is usually more complex and observation under a strobe light will show circular movement of the string.

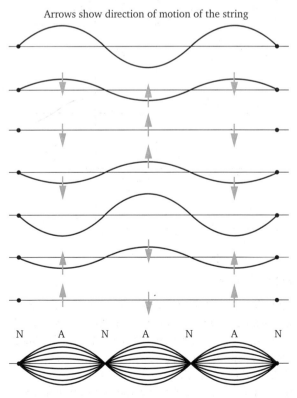

**Figure 2.66**

## Investigating the Factors that affect the Frequency

In the laboratory you should do experiments to check that the formula $f = \dfrac{1}{2l}\sqrt{\dfrac{T}{\mu}}$ for the frequency is correct in practice.

The equation suggests that experiments should show that:

$$f \propto \dfrac{1}{l} \text{ (when } T \text{ and } \mu \text{ are constant)}$$

$$f \propto \sqrt{T} \text{ (when } l \text{ and } \mu \text{ are constant)}$$

$$f \propto \dfrac{1}{\sqrt{\mu}} \text{ (when } T \text{ and } l \text{ are constant)}$$

The arrangement in Figure 2.67 can be used to investigate these relationships.

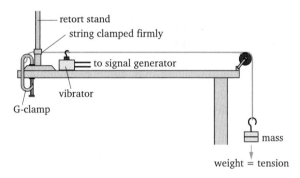

**Figure 2.67**

The weight of the mass attached to the string provides the tension. The signal generator connected to a vibrator is used to make the string vibrate in one of its modes. Only certain frequencies will produce stationary waves on the string.

## Testing $f \propto 1/l$

The same string is used throughout (so $\mu$ is constant). The weight is chosen so that a suitable length of string vibrates in its fundamental mode. The weight and hence the tension are kept constant throughout the experiment.

The frequency of the signal generator is adjusted to produce the fundamental mode. The frequency is determined from the scale of the signal generator or measured using an oscilloscope.

The length is recorded and changed. For each length the frequency that produces the fundamental vibration is determined.

Note that measuring the distance between two nodes for different vibrator frequencies may be more convenient than measuring the total length in this experiment.

A graph of $f$ against $1/l$ should be a straight line through the origin if the equation $f = \dfrac{1}{2l}\sqrt{\dfrac{T}{\mu}}$ is correct.

The other variables can be tested in a similar way. $T$ and $\mu$ can be varied in turn, keeping the other factors constant.

## Phenomena that Involve Stationary Transverse Waves

Bridges and buildings can vibrate transversely. Under certain conditions they can vibrate in one of their modes of vibration with a large amplitude, as shown in Figure 2.68. This could be caused by an earthquake, for example, and result in structural damage.

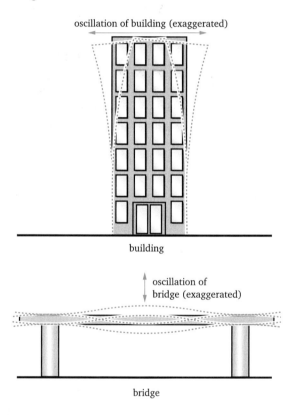

**Figure 2.68**

The building is only fixed at the bottom so this is a node. The top is free to vibrate and this can be thought of as an antinode. The height of the building is therefore $\dfrac{\lambda}{4}$.

## Looking Forward

In A2 you will learn that electrons in motion have wave properties. The ideas you have studied concerning the production of stationary waves are used to predict the discrete energies that an electron can have when confined inside an atom. These ideas are important in the modern view of the structure of atoms.

**Test your understanding**

1. A string resonates at a fundamental frequency of 200 Hz. The length of the string is 0.65 m.
   (a) Sketch a diagram to show this mode of vibration.
   (b) Calculate the frequency and wavelength of the third harmonic vibration for this string.
   (c) By how much would you need to increase the tension to double the fundamental frequency?
2. Determine the wavelength of the fifth harmonic of a string of length 0.80 m.
3. State what is meant by a *node*.
4. The second mode of vibration of a stretched string is at a frequency of 200 Hz. The length of the string is 0.80 m. The tension in the string is 50 N. Calculate the mass per unit length of the string.
5. A stretched string emits a note of 250 Hz when the tension is 36 N. Sketch a graph showing how the frequency of the fundamental of this string will vary with the tension.
6. (a) Two strings in a piano have the same tension and mass per unit length. One is twice the length of the other. The longer string emits a note at its a fundamental of 128 Hz. Determine the fundamental frequency of the other string.
   (b) How would the tension in the shorter string have to change for it to emit the same note as the longer string?
7. The frequency of vibration of the wire in the system in Figure 2.69 is 280 Hz.

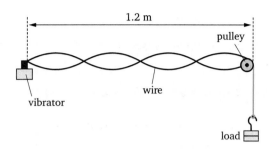

**Figure 2.69**

   (a) Calculate the speed of propagation of the wave along the wire.
   (b) Explain why the wire cannot vibrate at 56 Hz.
   (c) At what other frequencies below 280 Hz could it vibrate?

# Electromagnetic Waves

Visible light waves are one part of a very wide range of radiation all of the same type called **electromagnetic radiation**.

Unlike mechanical waves, which need a material medium to transfer the energy, electromagnetic waves can travel through a vacuum.

As the name implies, the energy is transferred by a variation of electric and magnetic fields. These field oscillations are transverse and are at right angles to one another.

Since the waves are transverse they can be polarized. Figure 2.70 represents a polarized electromagnetic wave travelling out of the paper.

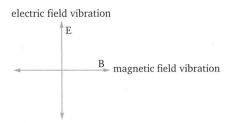

**Figure 2.70**

As energy travels through space the position of the maximum magnetic and electric fields move through space just like a transverse wave along a rope.

## Speed of Electromagnetic Radiation

The speed $c$ of electromagnetic radiation in free space (a vacuum) is $3.0 \times 10^8$ m s$^{-1}$ ($2.998 \times 10^8$ m s$^{-1}$). This is the same for all types of electromagnetic radiation.

The speed in media other than free space depends on the wavelength of the radiation. For example, blue light travels more slowly than red light in glass.

Television waves travel from the receiving aerial to the television through coaxial cable. The speed is about $\frac{2}{3}$ of the speed in air. This is because the wave has to travel through the polythene insulation that is between the central wire and copper screening.

As with all waves, the wavelength of electromagnetic radiation is related to the frequency of the radiation by the formula $v = f\lambda$. In this case $v = c$.

## Photons

Unlike the waves along a rope, which deliver energy continuously, electromagnetic wave energy is delivered by **photons**.

- A photon is a **wave 'packet'** that contains a certain amount of energy.
- The energy of a photon is proportional to the frequency of the wave, so higher frequency (shorter wavelength) waves have photons of higher energy.
- A photon of high energy transfers more energy when it interacts with matter. This is why photons of ultraviolet radiation or gamma radiation can cause serious damage to body cells whereas visible light does not.

The theory of wave-particle duality is covered in more detail in A2.

## The Electromagnetic Spectrum

This refers to the whole range of frequencies of electromagnetic waves. The name given to the different parts of the spectrum depends on how the waves were formed (i.e. their origin). The detection method depends on the photon energy and hence on the frequency and wavelength of the radiation.

The electromagnetic spectrum is continuous, with frequencies produced on Earth ranging from about $10^{-13}$ m for gamma radiation to about $5 \times 10^6$ m for waves generated by power lines.

Long-wavelength electromagnetic radiation ($\lambda \approx 3 \times 10^7$ m) has been detected in cosmic radiation from outer space.

Table 2.3 shows the types of electromagnetic radiation, their typical wavelengths $\lambda$, frequencies $f$ and photon energies, and their origin and method of detection.

Note that there is no distinct dividing line between the types (between X-rays and gamma rays, for example). The regions overlap and in the overlapping regions they are indistinguishable. The only difference is the way they are produced.

Details of some of these methods of production and detection and the associated physics are studied in A2.

## Table 2.3

| Type | $f$/Hz | $\lambda$/m | Typical photon energy (J) | Production method | Detection | Some uses |
|---|---|---|---|---|---|---|
| Gamma rays | $3 \times 10^{21}$ | $10^{-13}$ | $2 \times 10^{-12}$ | Radioactive decay of an excited nucleus | Photographic film Geiger tube Scintillation detectors | Information about the nucleus Medical diagnosis (radiography) Medical treatment of tumours (radiotherapy) Detection of flaws in metals |
| X-rays | $3 \times 10^{18}$ | $10^{-10}$ | $2 \times 10^{-15}$ | Inner electrons in atoms moving from one energy level to another Rapid deceleration of electrons during collisions in X-ray tubes | Photographic film Geiger tube | Radiography and radiotherapy (radiography) Crystallography Radio astronomy |
| Ultraviolet | $3 \times 10^{15}$ | $10^{-7}$ | $2 \times 10^{-18}$ | Inner and outer electrons in atoms moving from one energy level to another | Photographic film Photoelectric effect and photomultiplier | Industrial hardening of resins Identification of security ultra violet sensitive markings |
| Visible | $7.5 \times 10^{16}$ to $1.3 \times 10^{16}$ | $4 \times 10^{-7}$ to $7 \times 10^{-7}$ | $5 \times 10^{-17}$ to $8 \times 10^{-18}$ | Outer electrons in atoms moving from one energy level to another | Photographic film Effect on cells in retina | Sight Photosynthesis Fibre optic communication Chemical analysis by spectroscopy Laser technology Information from space in astronomy |
| Infrared | $3 \times 10^{12}$ | $10^{-4}$ | $2 \times 10^{-21}$ | Outer electrons in atoms moving from one energy level to another Atomic vibrations Molecular vibrations and rotations | Photographic film Thermopile | Heating Thermal detection of intruders in burglar alarms Infrared photography using satellites Information from space in astronomy |
| Microwaves | $1 \times 10^{10}$ | $3 \times 10^{-2}$ | $7 \times 10^{-24}$ | Klystron or magnetron valves | Diodes Heating effect | Satellite communication systems Cooking Radar, both on Earth and for astronomical purposes |
| UHF/VHF radio | $1 \times 10^{9}$ | $3 \times 10^{-1}$ | $7 \times 10^{-25}$ | Deceleration of electrons in aerials | Aerials | Television transmission VHF frequency modulated (FM) radio Mobile phones |
| Short/medium-wave radio | $1.5 \times 10^{6}$ | 200 m | $1 \times 10^{-27}$ | | | Commercial amplitude modulated (AM) radio |
| Long-wave radio | $3 \times 10^{4}$ | $1 \times 10^{4}$ | $2 \times 10^{-29}$ | | | Commercial AM radio |
| Power lines | 50 | $6 \times 10^{6}$ | $3 \times 10^{-32}$ | Transmission lines for electrical energy | Aerials (as interference) | |

## Emission Spectra

The light emitted from a body may give rise to:
- a continuous spectrum
- a line spectrum
- a band spectrum.

A **continuous emission spectrum**:
- is emitted by incandescent (glowing) solids and liquids
- depends on the temperature of the body
- contains all frequencies that are possible at the temperature of the body.

A filament, glowing red, contains all wavelengths equal to or longer than those in the red part of the visible spectrum. As the temperature rises, the atoms emit higher energy photons and the spectrum contains shorter and shorter wavelengths. Figure 2.71 shows the energy distribution of wavelengths for two temperatures.

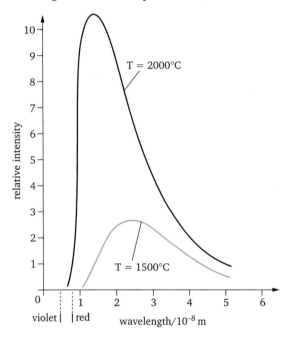

**Figure 2.71**

As the temperature increases, the colour of the source changes, becoming first red, then orange, yellow and finally white.

By examining the spectrum emitted by a hot body such as a star or furnace it is possible to determine its surface temperature. The surface temperature of the Sun has been estimated to be about 6000 K from its continuous spectrum.

A **line emission spectrum**:
- is emitted when atoms are excited by the discharge of electricity through a gas or by heating to a high temperature, for example in a Bunsen flame
- contains only photons of specific energy, i.e. light of specific wavelengths
- has a pattern that is characteristic of the atom that is producing the spectrum.

The wavelengths from a source can be precisely measured using diffraction gratings. By comparing the measured wavelengths with tables of wavelengths from known substances it is possible to determine which elements are emitting the light. This technique is used to analyse chemicals.

The **gamma ray spectra** emitted from excited nuclei provide evidence for the structure of the nuclei and show that neutrons and protons exist in well-defined energy levels in a nucleus.

In astronomy, emission spectra are examined in all parts of the electromagnetic spectrum (optical, microwave and X-ray). This provides information about the elements and processes that take place in stars.

A **band spectrum**:
- is produced by molecules rather than atoms
- contains groups of spectral lines
- enables analysis of the bonds between the atoms in a molecular structure.

## Absorption Spectra

When intense white light (light that has a continuous spectrum) is passed through a hot gas that contains atoms of a particular element, the resulting spectrum consists of a continuous spectrum crossed by dark lines or bands (Figure 2.72).

**Figure 2.72**

The dark lines (or bands) are due to wavelengths that have been reduced in intensity in passing through the gas. (The reasons for this will be studied in A2.)

The lines are characteristic of the atoms that are in the hot gas and correspond to those in the emission spectrum of the particular element.

The spectrum of the Sun is an absorption spectrum since the light emitted from its surface passes through the gases that surround the Sun. (These are called **Fraunhofer lines**.)

The elements in the gas surrounding the Sun can be determined from the absorption spectrum. Helium was identified in this way before it was extracted from air on Earth.

Figure 2.73 shows a method of producing an absorption spectrum in a laboratory.

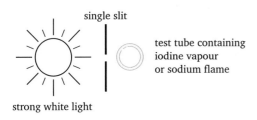
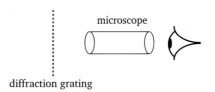

**Figure 2.73**

### Test your understanding

1. Place the following electromagnetic waves in order of increasing wavelength:
   red light, microwaves, green light, gamma radiation, ultraviolet radiation
2. Place the following types of electromagnetic radiation in order of increasing frequency:
   ultraviolet, X-rays, radio waves, blue light, infrared
3. (a) What is a photon?
   (b) Explain why radiation of different types has different effects on the human body.
4. A photon's energy is given by $6.6 \times 10^{-34} \times f$, where $f$ is the frequency of the radiation.
   (a) Write down an equation for the energy of a photon in terms of $c$, the speed of electromagnetic radiation ($3.0 \times 10^8$ m s$^{-1}$), and $\lambda$, the wavelength of the radiation.
   (b) Calculate the photon energies of the following electromagnetic radiations:
     (i) wavelength = 500 nm
     (ii) frequency = $2.5 \times 10^{10}$ Hz.
5. (a) State one use of electromagnetic radiation of the following wavelengths:
     (i) $2 \times 10^{-10}$ m
     (ii) 500 nm
     (iii) 3 cm.
   (b) State briefly how radiation of each of the following frequencies originates:
     (i) $5 \times 10^{21}$ Hz
     (ii) $5 \times 10^{16}$ Hz
     (iii) $1.5 \times 10^9$ Hz.
6. State the difference between an emission line spectrum and a band spectrum.

# Doppler Effect

The Doppler effect is the change in frequency observed when a source of a wave and an observer move towards or away from one another. An everyday example is the noticeable drop in pitch of the sound of a car horn or engine as a car passes a stationary observer.

When the source is moving toward the observer the observed frequency is higher than the frequency emitted by the source. When the source is moving away the frequency is lower.

### Source moving away from the observer

When a source emitting a frequency $f$ and an observer are both stationary, the $f$ waves occupy a distance $c$, where $c$ is the speed of the wave. The wavelength is $\frac{c}{f}$.

Because the source is moving away, the waves are stretched and will occupy a distance $c + v$, where $v$ is the speed of the source. This is illustrated in Figure 2.74.

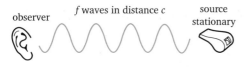

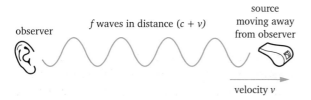

**Figure 2.74**

The observed wavelength is given by:

$$\lambda = \frac{c+v}{f}$$

The observed frequency is then:

$$f_{obs} = \frac{c}{\lambda} = \frac{c}{(c+v)} f$$

The frequency is therefore lower since $c < (c+v)$.

The change in frequency is $\Delta f = f - f_{obs}$:

$$\Delta f = f - \frac{c}{(c+v)} f$$

$$\Delta f = \frac{v}{(c+v)} f$$

The corresponding equation when the source moves toward the observer with the waves compressed into a smaller distance is:

$$\Delta f = \frac{v}{(c-v)} f$$

### Source speed $\ll$ speed of the wave

When this condition exists $(c + v)$ or $(c - v)$ is approximately equal to $c$.

The change in frequency is given to a good approximation by:

$$\Delta f = \frac{v}{c} f$$

or

$$\frac{\Delta f}{f} = \frac{v}{c}$$

In the case of sound this will only be approximately true for source speeds that are low. A speed of 30 m s$^{-1}$ produces an error of about 10% in the change of frequency when the approximate formula is used.

The approximate formula for the shift in wavelength, $\Delta \lambda$, is similar:

$$\frac{\Delta \lambda}{\lambda} = \frac{v}{c}$$

Note that you will only be required to use the approximate formula in examination questions.

## Uses of the Doppler effect

### Radar speed traps

Figure 2.75 illustrates how a radar speed trap might work. A transmitter emits microwaves, which reflect off a vehicle moving towards the transmitter.

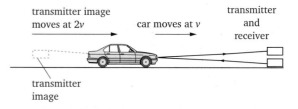

**Figure 2.75**

The vehicle behaves as a moving source. The frequency received back from the car is higher than that transmitted and the frequency received increases as the speed of the vehicle increases.

The radar device compares the outgoing and incoming waves and produces a readout showing the speed of the vehicle. The waves go in and out of phase, producing alternate loud and quiet sounds called beats. The frequency of the beats increases as the speed of the vehicle increases.

As far as the receiver is concerned, the waves appear to come from the image of the transmitter. This moves at twice the speed of the vehicle. The speed of the vehicle is therefore given by:

$$v = \frac{1}{2}\left(\frac{\Delta f}{f}c\right)$$

### Information about the body

A similar method is used to determine the rate of flow of blood in veins in the body. This can help diagnose heart disease or thrombosis, which reduces the blood flow rate due to a constriction in a vein.

Ultrasound of frequency 2.5–8 MHz is used to measure the blood flow rate. If the cross-sectional area of the vein is known, the flow rate can be determined.

### Information from space

The Doppler effect provides some evidence for an expanding universe and consequently for the big bang theory for the origin of the universe.

The frequencies of light in the line spectrum of atoms of an element in a galaxy that is moving away from Earth have a lower frequency (longer wavelength) than those of the light emitted from similar atoms on Earth. This is the **red shift**, so-called because the light frequency is shifted toward the red end of the optical spectrum. The reverse is true for galaxies approaching the Earth. Observations show that the red shift is observed more often than blue shifts, suggesting that the universe is expanding.

The difference between the actual and the expected frequency can be determined using diffraction grating techniques.

Using the approximate formula the speed at which a galaxy is moving away from our galaxy, the solar system, can be calculated. This is called the **recession speed**.

EXAMPLE ONE: A particular frequency emitted from a element in a galaxy is found to be $7.3 \times 10^{14}$ Hz. The same element on Earth emits light of frequency $7.6 \times 10^{14}$ Hz.

(a) Calculate the speed at which the galaxy is moving away from the Earth.

(b) What would be the measured frequency of the light if the galaxy had been approaching at the same speed?

(a) $$\frac{\Delta f}{f} = \frac{v}{c} \text{ so } v = \frac{\Delta f}{f}c$$

$$v = \frac{(7.6 - 7.3) \times 10^{14}}{7.6 \times 10^{14}} \times 3.0 \times 10^8$$

$$v = 1.1 \times 10^7 \text{ m s}^{-1}$$

(b) The frequency shift would be the same:

measured frequency = $(7.6 + 0.3) \times 10^{14}$ Hz
= $7.9 \times 10^{14}$ Hz

## The Hubble Law

Edwin Hubble produced his law from data derived from observations of the velocities of galaxies and their distances from the Earth. The velocities were determined by measurements of the red shift.

The Hubble law relates the **recession speed $v$** of a galaxy to its distance $d$ from the solar system.

The Hubble law states that:

$$v = Hd$$

where $H$ is the **Hubble constant**.

The Hubble constant is determined experimentally. The accepted value is 65 km s$^{-1}$ Mpc$^{-1}$, i.e. 65 kilometres per second per megaparsec (1 Mpc = $1 \times 10^6$ pc).

## The Parsec

The parsec (pc) is one of the units of distance used in astronomy. It is used because of the very large distances involved in astronomy and because of the way distance is measured using a process called trigonometric parallax. (This is not needed on the course but you might look up how distances are measured using this technique.)

1 parsec = $3.09 \times 10^{16}$ m
1 parsec = 3.26 light years

(You do not need to know these conversions. They will be given to you when needed.)

A **light year** is the distance travelled by light in one year.

The Hubble constant implies that the speed of a galaxy changes by 65 km s$^{-1}$ for every $3.09 \times 10^{22}$ m the galaxy is from the Earth, i.e. a star that is $3.09 \times 10^{22}$ m away is receding (moving away) at a speed of 65 km s$^{-1}$.

Although not very practicable, in more familiar units the Hubble constant is $2.1 \times 10^{-18}$ m s$^{-1}$ m$^{-1}$ (metres per second per metre away from the Earth).

Figure 2.76 shows how the recession speed varies with distance in Mpc. The edge of the universe is where $v = c$. This is at about 4600 Mpc or $1.42 \times 10^{26}$ m ($1.5 \times 10^{10}$ light years).

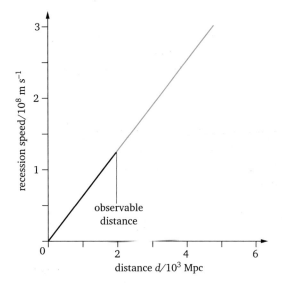

**Figure 2.76**

EXAMPLE TWO: How far away from the Earth is a galaxy that is receding at a speed of $1.1 \times 10^7$ m s$^{-1}$?

$$v = Hd$$

Changing the speed to km s$^{-1}$:

$$1.1 \times 10^4 = 65d$$
$$d = 169 \text{ Mpc}$$
$$= 169 \times 10^6 \times 3.1 \times 10^{16} \text{ m}$$
$$= 5.2 \times 10^{24} \text{ m}$$

## The Age of the Universe

At the edge of the universe galaxies move at speeds approaching the speed of light. Using the Hubble equation and the accepted value of the Hubble constant we can predict the edge of the universe to be $\dfrac{c}{H}$ parsecs, which is about 4600 Mpc away. This is $1.43 \times 10^{26}$ m.

Light must have travelled this distance since the big bang so on this theory the age of the universe is predicted to be $4.8 \times 10^{17}$ s or $1.5 \times 10^{10}$ years. Due to gravitational attraction slowing the expansion, the age is actually thought to be rather less than this, about $1.0 \times 10^{10}$ years.

**Test your understanding**

1. A police car has a siren that emits a frequency of 1500 Hz. The car moves toward an observer at a speed of 25 m s$^{-1}$. Sound travels at a speed of 340 m s$^{-1}$.
   (a) Use the accurate formula for the Doppler effect to determine the frequency heard:
      (i) as the car approaches a stationary observer
      (ii) as the car travels away from the observer.
   (b) Calculate the values in (a) using the approximation that assumes $c \gg v$.

2. The frequency of a train horn received by a stationary observer is 1430 Hz as it approaches and 1200 Hz as it passes the observer. The speed of sound in air is 340 m s$^{-1}$.
   (a) Calculate the speed of the train.
   (b) Calculate the frequency as the train moves directly away.

3. A stationary bat sends out a frequency of 50 kHz. The bat receives the signal back from a object moving away from it at a speed of 30 m s$^{-1}$. Calculate the frequency of the signal received by the bat.

4. Ultrasound of frequency 2.5 MHz is used to gather information about the inside of a body using the Doppler effect. The ultrasound travels at a speed of 1500 m s$^{-1}$. The maximum frequency observed in an examination of blood flow in an artery is 120 Hz higher than the emitted frequency. Calculate the maximum speed of the blood flow.

5. A galaxy is moving away from the Earth at a speed of $2.3 \times 10^7$ m s$^{-1}$. Hydrogen emits light of wavelength 410 nm. The speed of light is $3.0 \times 10^8$ m s$^{-1}$. Calculate the frequency of hydrogen light emission that you would expect to measure on Earth.

6. Use the data given in this section to calculate:
   (a) the distance in m that corresponds to the distance travelled by light in one year (1 light year)
   (b) the time taken for light to reach the Earth from a star that is 35 parsecs away
   (c) the distance from the Earth of a star that is receding from the Earth at 2500 km s$^{-1}$
   (d) the speed of a galaxy that is 100 Mpc away from the Earth.

7. The Hubble constant (65 km s$^{-1}$ Mpc$^{-1}$ may not be correct. How far away would the edge of the universe be if the Hubble constant were:
   (a) 20% smaller
   (b) 30% larger?

8. What would be the age of the universe, based on the Hubble constant, if the constant were:
   (a) 50 km s$^{-1}$ Mpc$^{-1}$
   (b) 80 km s$^{-1}$ Mpc$^{-1}$?

# Atomic Structure

The modern view of an atom is that it has:
- a diameter of about $10^{-10}$ m
- a very small **nucleus** (plural nuclei) of diameter about $10^{-15}$ m
- an electron cloud surrounding the nucleus.

The nucleus consists of **protons** and **neutrons**. These are both **nucleons**.

*Table 2.4*

| Particle | Mass/kg | Charge/C |
|---|---|---|
| Proton | $1.673 \times 10^{-27}$ | $+1.60 \times 10^{-19}$ |
| Neutron | $1.675 \times 10^{-27}$ | 0 |
| Electron | $9.11 \times 10^{-31}$ | $-1.60 \times 10^{-19}$ |

For most purposes at AS and A2 level the proton and neutron may be assumed to have equal masses of 1.67 (1.7) $\times 10^{-27}$ kg.

## Proton and Nucleon Numbers

The **proton number** $Z$ is the number of protons in a nucleus. This determines the element it is, e.g. hydrogen has $Z = 1$, helium has $Z = 2$, lithium has $Z = 3$, etc.

The **nucleon number** $A$ is the number of nucleons (protons + neutrons) in a nucleus.

The number of neutrons in a nucleus = $(A - Z)$.

The charge of a nucleus = $Z \times 1.6 \times 10^{-19}$ C.

The mass of a nucleus $\approx A \times 1.7 \times 10^{-27}$ kg.

The nuclear mass is not exactly the sum of the masses of the protons and neutrons that it contains but is very close. The reason why it is not exact will be discussed in A2.

Unless the atom is ionized, **the number of electrons in an atom is $Z$**, the same as the number of protons in the nucleus.

## Nuclear Nomenclature

This is the method used to describe a particular nuclear structure.

An atom X may be described as follows:
$$^A_Z X$$
where X is the chemical symbol for the element concerned.

$^1_1 H$ is hydrogen:
- $Z = 1$, so it has one proton in the nucleus
- $A = 1$, so there is one nucleon
- $(A - Z) = 0$, so this atom contains no neutrons in the nucleus.

$^4_2 He$ is helium:
- $Z = 2$, so there are two protons in the nucleus
- $A = 4$, so there are four nucleons
- $(A - Z) = 2$, so the nucleus contains two neutrons.

$^{238}_{92} U$ is uranium:
- there are 92 protons and $(238 - 92) = 146$ neutrons in its nucleus.

When the name or a symbol is given for an element the nucleon number is implied. When discussing atoms it is common to talk, for example, of lithium-7 or uranium-235.

However, to write nuclear equations (see later) the $Z$ number needs to be stated.

## Nuclides and Isotopes

These terms are both in common use to describe an atom that has a particular nuclear structure. The term 'isotope' is often used where the term 'nuclide' would be more appropriate.

A **nuclide** is simply a nucleus with a particular combination of protons and neutrons.

There are many nuclides. Some occur naturally in the Earth's crust and many more have been made in nuclear reactors and particle accelerators.

**Isotopes** are nuclides that have:
- **the same number of protons**
- **different numbers of neutrons**.

It follows that isotopes are all nuclei of the same element. Isotopes react chemically in the same way.

$^{28}_{14}Si$, $^{29}_{14}Si$ and $^{30}_{14}Si$ are all naturally occurring isotopes of silicon. They each have 14 protons in the nucleus. One has 14, one 15 and the other 16 neutrons in the nucleus.

A sample of naturally occurring silicon extracted from sand will contain each of these isotopes of silicon. About 92.2% is silicon-28, 4.7% is silicon-29 and 3.1% is silicon-30. These percentages are

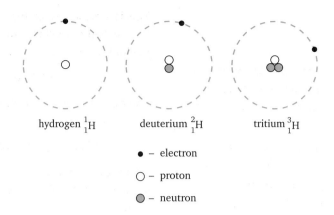

**Figure 2.77**

referred to as the natural abundances of the isotopes of silicon.

Figure 2.77 shows the three isotopes of hydrogen $^1_1H$, $^2_1H$ and $^3_1H$.

These are the only isotopes to have their own names. $^2_1H$ is called deuterium and $^3_1H$ is called tritium. The nucleus of deuterium is a deuteron (one proton and one neutron).

## Graph of Neutrons against Protons

Figure 2.78 shows how the number of neutrons in a nucleus varies with the proton number:
- for low $Z$ the number of neutrons is approximately equal to the number of protons
- as $Z$ increases the number of neutrons becomes much larger than the number of protons.

The extra neutrons are needed to counteract the repulsive force between the positively charged protons. The stable nuclides lie on or close to the blue line.

Nuclides to the right of the line are likely to be $\beta^+$ emitters and those to the left $\beta^-$ emitters.

### Test your understanding

1. Describe the atomic structure of gold, $^{197}_{79}Au$.
2. Calculate the approximate mass of:
   (a) a nucleus of uranium-235
   (b) an atom of hydrogen-3
   (c) an alpha particle.
3. A nucleus has a mass of $2.72 \times 10^{-26}$ kg and a charge of $1.28 \times 10^{-18}$ C.
   Use the data given in this section to determine:
   (a) the number of nucleons in the nucleus
   (b) the number of protons in the nucleus.
4. State what is meant when we say that two nuclides are isotopes.
5. Complete Table 2.5 to show the symbol for the isotope, proton number, the nucleon number, the neutron number and the number of electrons in each of the neutral atoms listed.

**Table 2.5**

| Symbol | $^{241}_{95}Am$ | Rn | Np | Pu | K | $^{60}Co$ |
|---|---|---|---|---|---|---|
| Proton number | | 86 | | | | 27 |
| Nucleon number | | 220 | 237 | 240 | | |
| Neutron number | | | 144 | | 21 | |
| Electron number | | | | 94 | 19 | |

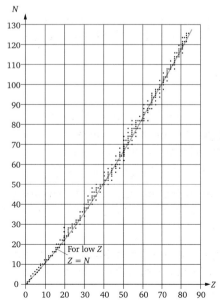

**Figure 2.78**

# Radioactivity

## What is radioactivity?

The nuclei of some atoms are unstable and undergo **spontaneous changes**. A spontaneous change takes place without any outside stimulus to make it happen.

The nuclei that emit such radiation are called radioactive nuclides or radioactive isotopes of the element. A radioactive isotope emits energetic particles or energy in the form of electromagnetic radiation to become more stable. There are changes in the nucleus as a result of the emission.

The three types of emission from naturally occurring radioactive nuclei are:
- **alpha (α) particles**
- **beta⁻ (β⁻) particles** (beta-minus particles)
- **gamma (γ) radiation**.

Radioactive nuclides made in particle accelerators and nuclear reactors may also decay by emitting a **beta⁺ (β⁺) particle**, called a **positron**.

The nucleus emitting the radiation is called the **parent nucleus**. The nucleus that is left after the emission is called the **daughter nucleus**.

## Alpha Emission

An alpha particle consists of two protons and two neutrons and is therefore the same as a helium-4 nucleus.

An alpha particle has a:
- charge of $+2 \times 1.6 \times 10^{-19}$ (i.e. $3.2 \times 10^{-19}$ C)
- mass of $4 \times 1.7 \times 10^{-27}$ (i.e. $6.8 \times 10^{-27}$ kg).

The sign of the charge on an alpha particle can be determined by observing the direction of its deflection in a magnetic field. An alpha particle and a beta⁻ particle deflect in opposite directions. The fact that much larger magnetic fields are needed to produce noticeable deflection of alpha particles is evidence that its mass is considerably larger than that of an electron.

An alpha particle carries energy in the form of kinetic energy. The kinetic energy of an alpha particle is about $8 \times 10^{-13}$ J. All the alpha particles produced by a given nuclide have the same energy.

### Changes in the nucleus following alpha emission

When alpha emission occurs:
- the proton number $Z$ decreases by 2
- the nucleon number $A$ decreases by 4.

This is represented by the nuclear equation in which P is the parent nucleus and D is the daughter nucleus:

$$^{A}_{Z}P \rightarrow {}^{A-4}_{Z-2}D + {}^{4}_{2}\alpha + \text{kinetic energy}$$

The following equation shows the decay of radon-220 into polonium:

$$^{220}_{84}Rn \rightarrow {}^{216}_{82}Po + {}^{4}_{2}\alpha + 1.0 \times 10^{-12} \text{ J}$$

Notice that in any nuclear equation:
- the sum of the $Z$ numbers on each side must be equal
- the sum of the $A$ numbers on each side must be equal
- the number of nucleons (the baryon number) is conserved.

### The electron-volt

The energy of the emissions from radioactive nuclei is usually given in electron-volts (eV):

$$1 \text{ eV} = 1.6 \times 10^{-19} \text{ J}$$

(You will not need to recall this information in AS. If required it will be provided in questions.)

A typical alpha particle therefore has an energy equal to:

$$\frac{8 \times 10^{-13}}{1.6 \times 10^{-19}} = 5 \times 10^{6} \text{ eV or 5 MeV}$$

### Speed of an alpha particle

The speed of an alpha particle can be calculated using $\frac{1}{2}mv^2$.

The mass of an alpha particle is $4 \times 1.7 \times 10^{-27}$ kg = $6.8 \times 10^{-27}$ kg.

Using the typical energy of $8 \times 10^{-13}$ J gives an alpha particle speed of $1.5 \times 10^{7}$ m s⁻¹ when it is emitted.

### Recoil nucleus

When an alpha particle is emitted some of the energy appears as the kinetic energy of the nucleus as it recoils. This is the same effect as when a cannon recoils when a cannon ball is fired.

Because the nucleus usually has much more mass than the alpha particle, most of the energy is carried by the alpha particle.

In the decay of radon shown in Figure 2.79 the alpha particle carries over 98% of the energy lost by the nucleus.

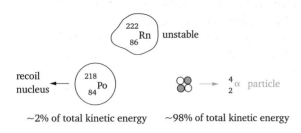

**Figure 2.79**

## Beta⁻ Emission

A beta⁻ particle is a fast-moving electron that is emitted from the nucleus and therefore carries a charge of $-1.6 \times 10^{-19}$ C. This is illustrated by the fact that when they are slowed down, beta⁻ particles behave in the same way in electric and magnetic fields as electrons produced by other methods (e.g. hot filaments). Like alpha particles, the energy is carried in the form of kinetic energy. Unlike alpha particles, the energy can take any value up to a maximum value (see below). The typical maximum energy of a beta⁻ particle from a given nucleus is $2 \times 10^{-13}$ J.

Figure 2.80 is a beta⁻ particle spectrum and shows the relative number of beta⁻ particles with a given energy from potassium-40.

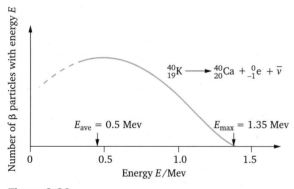

**Figure 2.80**

The fact that beta⁻ particles from a particular nuclide have a range of possible energies is evidence for the existence of neutrinos (see later). Beta⁻ emission is accompanied by the emission of an anti-neutrino, $\bar{v}$. The line across the top of this symbol signifies that it is an antiparticle.

### Changes in the nucleus following beta⁻ emission

When a nucleus decays by beta⁻ emission:
- the proton number $Z$ increases by 1
- the nucleon number $A$ remains the same.

This is represented by the following nuclear equation in which P is the parent nucleus and D is the daughter nucleus:

$$^{A}_{Z}P \Rightarrow \,^{A}_{Z+1}D + \,^{0}_{-1}\beta + \,^{0}_{0}\bar{v} + \text{kinetic energy}$$

The following equation shows the decay of potassium-40 into calcium-40:

$$^{40}_{19}K \Rightarrow \,^{40}_{20}Ca + \,^{0}_{-1}\beta^{-} + \,^{0}_{0}\bar{v} + 2.2 \times 10^{-13} \text{ J}$$

### Speed of a beta particle

An electron has a mass of $9.1 \times 10^{-31}$ kg. Assuming a typical energy of $2 \times 10^{-13}$ J and using the usual equation for kinetic energy gives a speed of $6.6 \times 10^{8}$ m s⁻¹.

This exceeds the speed of light and so is impossible. Because the electrons are moving close to the speed of light when they are emitted, their energy appears as increased mass instead of higher speed. At the high speeds involved Einstein's relativity theory has to be applied.

## Positron Emission

When a nucleus decays by positron (beta⁺) emission:
- the proton number $Z$ decreases by 1
- the nucleon number $A$ remains the same
- a neutrino $v$ is emitted.

This is represented by the following nuclear equation, in which P is the parent nucleus and D is the daughter nucleus:

$$^{A}_{Z}P \Rightarrow \,^{A}_{Z-1}D + \,^{0}_{1}\beta^{+} + \,^{0}_{0}v + \text{kinetic energy}$$

The following equation shows the decay of aluminium-25 into magnesium-25:

$$^{25}_{13}Al \Rightarrow \,^{25}_{12}Mg + \,^{0}_{1}\beta^{+} + \,^{0}_{0}v + 5.1 \times 10^{-13} \text{ J (3.2 MeV)}$$

Most nuclei that emit positrons do not exist for very long after they are formed.

## Gamma Radiation

Gamma radiation is emitted in the form of photons of electromagnetic radiation. Like all electromagnetic radiation they travel at $3.0 \times 10^{8}$ m s⁻¹ in free space.

Gamma-ray photons:
- are emitted at the same time or very soon after an alpha or a beta decay
- result from an excited nucleus readjusting the nucleons into a more stable arrangement.

A gamma-ray photon has a typical energy of about $5 \times 10^{-14}$ J (about 300 keV). Such radiation has a frequency of about $7.6 \times 10^{20}$ Hz and wavelength about $4 \times 10^{-12}$ m.

### Changes in the nucleus following gamma emission

Emission of a gamma-ray photon has **no effect** on the proton number or the nucleon number.

The only change is a rearrangement of the nucleons to produce a more stable nucleus.

## Evidence for the Existence of Neutrinos

Neutrinos and antineutrinos are particles with negligible rest mass. However, they have energy and momentum when moving. Although there are vast numbers of neutrinos arriving at the Earth from the beta decays all over the universe, they do not interact often with nuclei and are very difficult particles to detect.

Neutrinos were thought to exist because observations of beta decays appeared to break the well-established laws of conservation of energy and conservation of momentum. The evidence is given in more detail below.

It is now known that other particle physics laws would also be broken without the existence of the neutrino (see pages 107–109).

### Energy considerations

When a particular nuclide undergoes decay by beta⁻ emission the final nucleus is identical in each case so it must lose the same energy overall.

The beta⁻ particles emitted have a range of energies (see Figure 2.80) and the total energy of the recoil nucleus and the beta⁻ particle also varies. This suggests that nucleus must therefore be losing energy in some other form.

The energy of the antineutrino in a given decay accounts for the difference between the maximum energy available for that decay and that of the beta⁻ particle and recoil nucleus. The energy of the beta⁻ particle emitted by a nucleus depends on how much energy is given to an antineutrino.

### Momentum considerations

Figure 2.81 shows the direction of motion of the beta⁻ particle and the recoil nucleus in a typical beta⁻ decay.

**Figure 2.81**

Application of the principle of conservation of momentum (studied in A2) shows that if only two particles are involved, the recoil nucleus should **always** move in the opposite direction to that of the beta⁻ particle. The only way momentum can be conserved in the situation shown in Figure 2.81 is for another particle (an antineutrino) to be emitted in the direction shown by the blue line in Figure 2.82.

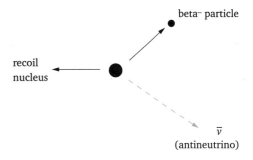

**Figure 2.82**

The magnitude and direction of the momentum of the antineutrino ensures that the conservation law holds.

**Test your understanding**

1. Write down the symbols for the following particles, including their mass (nucleon) numbers and their charge (proton) numbers: proton, electron, alpha particle, neutron, gamma ray, photon, neutrino
2. $^{206}_{82}Pb$ is formed when a radioactive atom decays by alpha emission. Determine the proton number and nucleon number of the atom that has decayed.
3. In the actinium series uranium-235 ($^{235}_{92}U$) decays by the emission of five alpha particles and two beta⁻ particles. Determine the proton and nucleon number of the nuclide that is formed when these decays have taken place.
4. Write the full nuclear equations for the following:
   (a) a free neutron decaying by beta⁻ decay
   (b) $^{11}_{6}C$ decaying by $\beta^+$ emission into boron
   (c) radium-226 (Ra-226) decaying by alpha emission into radon (Rn). Radon has a proton number of 86.
5. A nucleus of iodine (I) decays emitting an electron antineutrino.
   (a) What other particle is emitted?
   (b) Describe briefly the evidence for the emission of two particles in such decays.
   (c) The nucleus that is formed is Xe-129 ($^{129}_{54}Xe$). Write down the nuclear equation for the decay.
6. The speeds of an alpha particle and the recoil nucleus are inversely proportional to their masses. An alpha particle is emitted from U-235 with an energy of $7.3 \times 10^{-13}$ J. The mass of a proton is $1.7 \times 10^{-27}$ kg.
   Calculate:
   (a) the speed of the alpha particle
   (b) the speed of the recoil nucleus
   (c) the energy of the recoil nucleus
   (d) the percentage of the total energy carried by the recoil nucleus.

## Ionization

Ionization is the process of adding or removing one or more electrons from an atom or molecule. This leaves the particle with a net positive or negative charge. Figure 2.83 shows the ionization of an oxygen-16 atom.

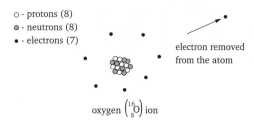

**Figure 2.83**

The energy required to ionize a molecule of oxygen or nitrogen in air is about $5 \times 10^{-18}$ J (30 eV). To ionize an atom of lead takes about $1.2 \times 10^{-18}$ J.

### Ionizing radiation

Alpha and beta particles are charged particles. Due to repulsion of like charges or attraction of unlike charges, they can transfer energy to an electron and produce ionization. Photons of gamma radiation can produce ionization when they collide with electrons in atoms.

Alpha and beta particles are both **strongly ionizing**. They produce many ions in a relatively short distance as they travel in a material.

Gamma radiation is weakly ionizing because the photons have no charge. Gamma radiation produces fewer ions over a given distance but as a result it can produce ions a long way from the source.

### Penetrating power

When an ion is produced the kinetic energy of an alpha or beta particle falls. This will continue until the particle transfers all its kinetic energy to electrons and stops. The penetration of a particle depends on:

- the type of particle – alpha particles carry more charge and mass than beta particles so produce ions more often
- the initial energy of the particle – the greater the energy the further they travel before they stop (larger range)
- the energy required to produce an ion – when less energy is needed to ionize an atom the range is larger
- the frequency of collisions – when atoms are far apart (as in gases) the range is larger than when they are closely packed (as in solids).

**EXAMPLE ONE:** An alpha particle is emitted with a kinetic energy of $6.0 \times 10^{-13}$ J. It travels 35 mm in a gas. The energy needed to ionize an atom of gas is $4.0 \times 10^{-18}$ J.
(a) How many ions can the particle produce?
(b) How many ions would you expect per millimetre?

(a) Number of ions =
$$\frac{\text{initial kinetic energy}}{\text{energy to produce one ion}} = \frac{6.0 \times 10^{-13}}{4.0 \times 10^{-18}} = 1.5 \times 10^5$$

(b) Number of ions per millimetre =
$$\frac{\text{total number of ions}}{\text{distance travelled}} = \frac{1.5 \times 10^5}{35} = 4300$$

In fact there would be more ions near the end of its range because as an alpha particle moves more slowly it stays longer in each millimetre and produces more ions in each millimetre of its range.

## Typical Ranges of Radiation

Alpha particles:
- travel between 30 and 100 mm in air at atmospheric pressure
- are stopped by a thin sheet of paper
- produce many ions close to the source so radioactive dust is particularly dangerous if breathed in or ingested.

Beta particles:
- travel between 0.2 and 0.5 m in air at atmospheric pressure
- are stopped by a few millimetres of aluminium
- can penetrate up to 20 mm in human flesh.

Gamma radiation:
- produces very few ions when travelling through air
- obeys an inverse square law (see later)
- is reduced to safe levels only after passing through several centimetres of lead or even thicker blocks of concrete.

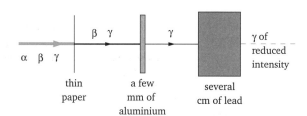

**Figure 2.84**

### Test your understanding

1. The energy of an alpha particle is 5.0 MeV. The energy needed to produce an ion in hydrogen is 13.6 eV.
   (a) How many ions can the alpha particle produce in hydrogen?
   (b) Explain how would you expect the range of the alpha particle to change when the pressure of the hydrogen is increased by compressing it.
2. (a) Explain what is meant by ionization.
   (b) Explain why a proton produces more ions per millimetre than a neutron with the same kinetic energy.
3. Explain why alpha particles produce more ions per millimetre in air than a beta particle that has the same initial energy.
4. Figure 2.85 shows how the number of ions varies with distance from a source emitting alpha particles.

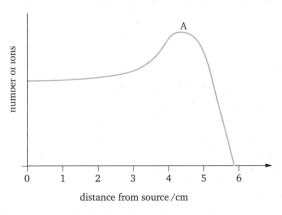

**Figure 2.85**

   (a) Explain why there is an increase in the number of ions per millimetre at A.
   (b) Determine the range of the alpha particles.
   (c) Copy Figure 2.85 and draw on it a second graph for an alpha particle from the same source travelling through a denser gas.

## Activity

The activity of a source is the number of disintegrations that take place per second in the source. Activity is measured in becquerels:

1 becquerel (Bq) = 1 disintegration per second

The sources used in a typical A-level laboratory are all sealed so that no atoms of the radioactive source can escape and contaminate the laboratory. Their activities range from about $3 \times 10^4$ Bq to $15 \times 10^4$ Bq.

## Count Rate

This is the number of counts recorded per second by a counter connected to a detector such as a Geiger–Müller tube shown in blue in Figure 2.86:

$$\text{count rate} = \frac{\text{total number of counts recorded}}{\text{time taken}}$$

The units are counts per second ($s^{-1}$).

Note that the count rate is not the same as activity. As shown in Figure 2.86, only a small fraction of the radiation from a radioactive source passes through the detector and not all of this is counted.

## Background Radiation

A detector records a count rate even when there are no obvious sources of radiation present. The counts, or clicks if an audible detector is used, are due to **background radiation** that exists in the environment.

The count rate due to background radiation is random at any time, i.e. the counts do not occur at fixed time intervals. It can also vary during the day owing to varying cosmic activity.

Background radiation is caused by:
- naturally occurring radioactive materials such as uranium in rocks and radon gas in the atmosphere

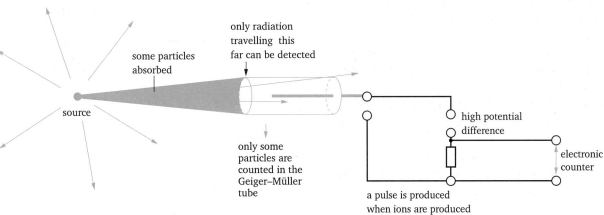

**Figure 2.86**

- radioactive carbon-14 in plants and animals, including ourselves
- particles from space (cosmic radiation)
- radioactive fallout from nuclear tests and nuclear accidents
- equipment such as X-ray machines and smoke detectors.

When carrying out experiments, the background count rate has to be measured using the same voltage setting that is used in the main experiment. The count rate due to a source is equal to the observed count rate minus the background count rate.

## Randomness of Radioactive Decay

The decay of radioactive nuclei is a random event. This means:
- it is not possible to tell when any particular nucleus will decay
- all nuclei of a given nuclide have the same chance of decaying at any time.

## The Decay Equation

The probability of a nucleus decaying in a given time interval is called the **decay constant** $\lambda$. The SI unit is the chance of decay per second.

The value of $\lambda$ depends on the nucleus and can have a wide range of values. For example:

$\lambda = 4.5 \times 10^{-17}$ s$^{-1}$    for uranium-235

$\lambda = 0.019$ s$^{-1}$    for radon

This means that a uranium-235 nucleus has a 4.5 in $1 \times 10^{17}$ chance of decaying each second. This is more usually stated as a 1 in $2.2 \times 10^{16}$ chance of decaying each second (since $\lambda = \dfrac{1}{2.2 \times 10^{16}}$).
For radon the chance is 1 in 53.

As all nuclei of a given nuclide have the same chance of decaying, the number decaying per second (the activity $A$) is given by:

$$A = \lambda N$$

where $N$ is the number of radioactive nuclei of the nuclide present.

Data books usually give the half-life ($T_{\frac{1}{2}}$) rather than the decay constant for a nucleus. These two parameters are related by the equation:

$$T_{\frac{1}{2}} = \dfrac{0.69}{\lambda} \text{ or } \lambda = \dfrac{0.69}{T_{\frac{1}{2}}}$$

EXAMPLE TWO: One gram of uranium-235 contains $2.6 \times 10^{21}$ atoms.
(a) What is the activity of 1 g of uranium-235?
(b) What mass of uranium-235 would have an activity of 1 Bq?

(a) $A = 4.5 \times 10^{-17} \times 2.6 \times 10^{21} = 11\,700$ Bq

(b) The number of radioactive atoms $N$ with an activity of 1 Bq is given by:

$$1 = 4.5 \times 10^{-17} \times N$$

$$N = \dfrac{1}{4.7 \times 10^{-17}} = 2.2 \times 10^{16}$$

$$\text{mass} = \dfrac{2.2 \times 10^{16}}{6.0 \times 10^{21}} = 3.7 \times 10^{-5} \text{ g or } 37\mu g$$

Notice that the activity gives the decrease in the number of radioactive atoms each second. This means that in the next second there will be fewer atoms to decay so the activity will decrease.

## Half-life

The **half-life** is the time taken for the activity of a particular nuclide to become half its initial value. Since $A \propto N$ the number of radioactive nuclei of the nuclide will also become half the initial value.

Graphs of $A$ against time $t$ and of $N$ against $t$ are shown in Figure 2.87. The initial activity is $N_0$ and the half-life is $T_{\frac{1}{2}}$.

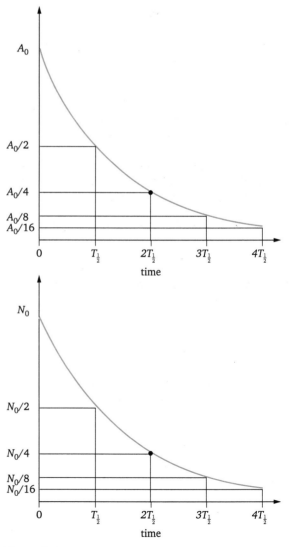

**Figure 2.87**

The number of radioactive nuclei left after

$$T_{\frac{1}{2}} \text{ is } \frac{1}{2} \times N_0 = \frac{N_0}{2}$$

$$2T_{\frac{1}{2}} \text{ is } \frac{1}{2} \times \frac{1}{2} \times N_0 = \frac{N_0}{4}$$

$$3T_{\frac{1}{2}} \text{ is } \frac{1}{2} \times \frac{1}{2} \times \frac{1}{2} \times N_0 = \frac{N_0}{8}$$

After $n$ half-lives there are $\left(\frac{1}{2}\right)^n \times N_0 = \frac{N_0}{2^n}$ radioactive atoms remaining.

The values of $A$ and $N$ at any given time may be read from a graph.

(In A2 you will learn how to calculate values of $A$ and $N$ at any given time.)

## Exponential Changes

The graphs in Figure 2.87 show exponential decay. This occurs because $A = \lambda N$.

The graphs in Figure 2.87 have the same shape because $A$ is proportional to $N$.

Note that any change where the rate of decrease is proportional to amount present at any time will give a graph of this shape.

## Practical Graphs

There are few sources that are suitable for measuring half-life in an AS laboratory. One nuclide that may be used is protactinium, Pa-234.

Protactinium-234 is a beta emitter that is formed from thorium-234, which in turn is produced from uranium-238. The decay series is:

$$\begin{array}{ccccc} {}^{238}_{92}\text{U} & \Rightarrow & {}^{234}_{90}\text{Th} & \Rightarrow & {}^{234}_{91}\text{Pa} & \Rightarrow & {}^{234}_{92}\text{U} \\ \downarrow & & \downarrow & & \downarrow \\ \alpha & & \beta^- & & \beta^- \end{array}$$

The measuring arrangement used is shown schematically in Figure 2.88.

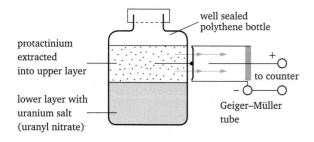

**Figure 2.88**

Normally all the radioactive isotopes are in the lower liquid layer, which is unable to mix permanently with the upper layer. The container is shaken so that the liquids mix. They are then allowed to separate again. Only the protactinium remains in the top layer.

Beta⁻ particles emitted by the protactinium can pass through the thin polythene container and are monitored using the Geiger–Müller tube and counter.

### General shape

The graph that is obtained practically is count rate against time. Because the count rate is proportional to the activity this will have the same general shape as the A–t and N–t graphs.

### Randomness

Usually the activity of the source used and therefore the count rates are relatively low and the randomness of the emission becomes evident. The points do not lie on a smooth curve.

The randomness is evident in Figure 2.89, which shows typical results against time for the protactinium experiment.

The blue line shows the raw data, which has not been correct for background radiation. The black line shows the result of applying the correction and smoothing the data. Each count rate has to be reduced by the same amount.

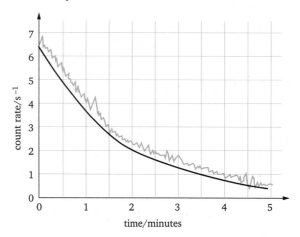

**Figure 2.89**

You should be able to see that the time taken for the count rate to halve is different for the two lines. Using the uncorrected graph will give a value for half-life that is longer than the correct value.

## Half-life of Long Half-life Sources

The half-life of uranium-238 is $4.51 \times 10^9$ years. Clearly this cannot be determined by measuring changes in activity. There is no real change in the measured activity of such a source in a person's lifetime.

One way of measuring the half-life is to extract a known mass of the isotope. The number of radioactive atoms can be determined from this mass (235 g of uranium-235 contains $6.0 \times 10^{23}$ atoms, as does 4 g of helium-4 and 12 g of carbon-12).

The total activity is measured (i.e. the total count rate, not just a sample count rate). The decay constant can be found using $A = \lambda N$ and the half-life from

$$T_{\frac{1}{2}} = \frac{0.69}{\lambda}$$

## Radiocarbon Dating

Because wood contains carbon, the decay of radioactive carbon-14 enables the determination of the approximate age of ancient wooden artefacts.

Radioactive carbon-14 is created in the atmosphere by neutrons in cosmic radiation:

$$^1_0n + ^{14}_7N \Rightarrow ^{14}_6C + ^1_1p$$

A small percentage of carbon in the atmosphere is carbon-14. Most is carbon-12. As it grows, a tree takes in carbon from the atmosphere and so is slightly radioactive. When the tree is used to make an artefact (e.g. a boat) it stops taking in carbon. The carbon-14 decays with a half-life of 5570 years.

The activity of a given mass of carbon taken from an old artefact (e.g. an old boat) is compared with that of a similar mass of carbon from a recently cut tree. Assuming that the proportion of radioactive carbon in the atmosphere has not changed over time, the age of the artefact can be calculated.

Figure 2.90 shows the decay graph for a given mass of carbon. The activity plotted on the y-axis is expressed as a percentage of the original activity. After 5600 years the activity is half (i.e. 50%) the original activity.

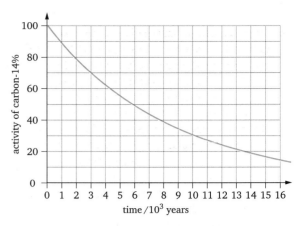

**Figure 2.90**

For example, the activity for one artefact is 70% of that of fresh carbon. The approximate age of the artefact, read from the graph, is therefore 3000 years.

## Energy Sources

As a source decays it releases energy. This can be tapped to provide the thermal energy for an electrical power supply. Typical uses are supplies for spacecraft and for heart pacemakers.

The source needs to have:
- a suitable half-life so that it can provide energy at a reasonably constant rate for the time it is to be used
- a suitable emission that can be easily absorbed to raise the temperature sufficiently
- no penetrating emissions that could harm people or electronic equipment.

### Test your understanding

1. How does the value for the half-life of protactinium from the data in Figure 2.89 compare with the accepted value of 70 s?
2. Complete Table 2.6.

**Table 2.6**

| Source | Mg-23 | Sr-90 | U-238 | Pa-234 |
|---|---|---|---|---|
| Half-life | 12 s | | $4.5 \times 10^9$ years | |
| Decay constant $\lambda/s^{-1}$ | | $7.8 \times 10^{-10}$ | | |
| Activity $A$/Bq | 20 | | 45 | $2.0 \times 10^4$ |
| Number of radioactive nuclei | | $2.5 \times 10^{20}$ | | $2.1 \times 10^6$ |

3. A radioactive nuclide has a half-life of 30 minutes. It initially contains $1.6 \times 10^{20}$ radioactive atoms.
   (a) Sketch a graph showing how the activity of the nuclide varies over the first 2 hours.
   (b) Calculate the decay constant of the nuclide.
   (c) Calculate the initial activity of the nuclide.

4  The ratio of radioactive carbon-14 to non-radioactive carbon in living matter is $1.3 \times 10^{-12}$.
   (a) 14 g of radioactive carbon-14 contains $6 \times 10^{23}$ atoms. Determine the mass of carbon-14 in 500 g of carbon from the bones of an animal that has recently died.
   (b) What is the activity of 500 g of bone taken from an animal that has recently died?
   (c) 500 g of carbon from some ancient bones has an activity of 26 Bq. Estimate the age of the old bones. (The decay constant of carbon-14 is $3.8 \times 10^{-12}$ s$^{-1}$.)

5  The half-life of potassium-42 is 12 h.
   (a) Calculate:
       (i) the decay constant in h$^{-1}$
       (ii) the decay constant in s$^{-1}$.
   (b) What percentage of the original radioactive potassium would you expect to be present:
       (i) after 12 h
       (ii) after 24 h?

6  Thorium-234 has a half-life of 24 days. A sample initially contains $6.0 \times 10^{12}$ atoms.
   (a) Calculate the decay constant of thorium in s$^{-1}$.
   (b) Calculate the initial activity of the thorium-23.
   (c) Plot a graph for the decay and determine:
       (i) the number of radioactive atoms present after 12 days
       (ii) the time at which the activity has fallen to 25% of the original activity.

7  Iodine-131 is used to treat disorders of the thyroid gland. A nucleus of iodine-131 ($^{131}_{53}$I) decays to xenon by beta$^-$ emission. Figure 2.91 shows how the activity of a sample of the isotope varied with time after being injected into a patient.

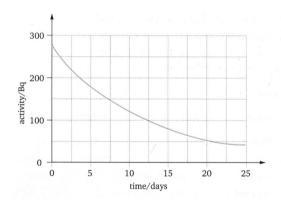

**Figure 2.91**

   (a) State what is meant by *half-life*.
   (b) State and explain **two** properties of radioactive isotopes that make them useful in medical treatment.
   (c) (i) What quantity does the area under the graph represent?
       (ii) Use the graph to determine the half-life of iodine-131.
       (iii) Calculate the decay constant of iodine-131 in s$^{-1}$.
   (d) (i) Determine the number of radioactive atoms that were introduced into the patient.
       (ii) Sketch a graph to show how the number of atoms of xenon varies with time following injection into the patient.
   (e) Describe briefly how you would demonstrate that the decay of iodine is a random process.

8  Sodium-24 ($^{24}_{11}$Na) is a beta$^-$ emitter that is used in hospitals as a tracer. Figure 2.92 shows how the activity of a sample of sodium-24 varies with time.

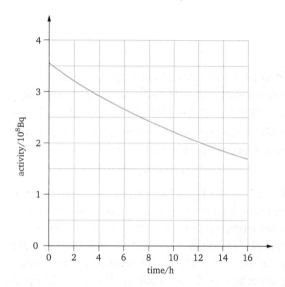

**Figure 2.92**

   (a) (i) What is meant by an activity of 1 Bq?
       (ii) Explain briefly how an isotope such as sodium-24 may be used as a tracer.
       (iii) Why is a beta or gamma emitter preferred to an alpha emitter for such purposes?
   (b) Sodium-24 is manufactured at Harwell, which is an 8 hour drive (allowing for traffic) from a hospital. On arrival at the hospital the source must have an activity of no less than $2.0 \times 10^8$ Bq.
       Determine:
       (i) the minimum activity of the source when it leaves Harwell
       (ii) the minimum number of radioactive iodine atoms present when the source arrives at the hospital.

## Detecting Radiation

### Geiger–Müller tube and counter

In an A-level laboratory the usual method of detecting radiation is to use a Geiger–Müller tube connected to an electronic counter, as shown in Figure 2.86.

When an alpha or beta particle or a gamma ray photon enters the tube ionization occurs and a pulse of current causes a count on the counter. (It is not essential for you to know any details of the operation of the counter.)

Not all radiation arriving at the Geiger–Müller tube causes a count because:
- some alpha particles are absorbed by the thin mica window
- a gamma-ray photon may not produce any ions in the gas (since a collision with a gas atom is rare)
- a particle or photon that arrives after another cannot be detected unless the tube has recovered from the previous count.

Although it is not a true value of the intensity of the radiation, the count rate can be assumed to be proportional to the intensity.

### Other detectors

Some other methods of detecting radiation are:
- photographic film
- ionization chambers
- cloud chambers
- bubble chambers
- spark detectors
- solid-state detectors.

**Photographic film** is used to check that those who work with radiation have not been exposed to more radiation than is permitted under health and safety rules. Workers wear a badge consisting of film behind absorbers of different thickness (Figure 2.93).

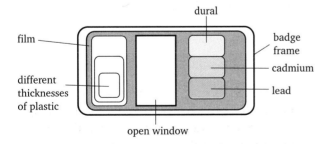

**Figure 2.93**

The darkness of the film depends on the radiation received. Darkening behind a thicker layer or a denser absorber signifies exposure to more penetrating radiation.

## Conducting Experiments with Radiation

Although you will only deal with low-activity sealed sources you must observe the following safety rules when conducting experiments:
- check the background radiation before and after experiments
- take sources from the radioactivity store when they are needed and return them immediately you have finished the experiment
- always handle the sources with tongs
- keep the sources as far away from you and others as possible when handling them and when setting up your experiment
- use shielding for protection
- do not eat or drink in the laboratory
- wash your hands after working with sources.

### Radiation absorption experiments

An arrangement for investigating the absorption of radiation in an A-level physics laboratory is shown in Figure 2.94. The arrangement is sometimes in the form of a metal 'castle', which provides screening between the experimenter and the source.

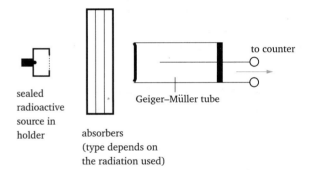

**Figure 2.94**

In each experiment it is necessary to first measure the background count rate when no source is present.

### Alpha particle emitter experiments

- Determine the count rate with the source close to the Geiger–Müller tube. (Some alpha particles should be able to enter through the thin mica window at the front of the tube.)
- Put a thin sheet of paper between the source and the detector. The count rate should drop considerably.
- Without the paper in place, increase the distance between the source and the detector by a few millimetres. Record the count rate for each setting. The count rate should fall and become very low when the separation is about 40 mm. This is an estimate of the range of the alpha particles.
- The range increases for sources that emit higher energy α particles.

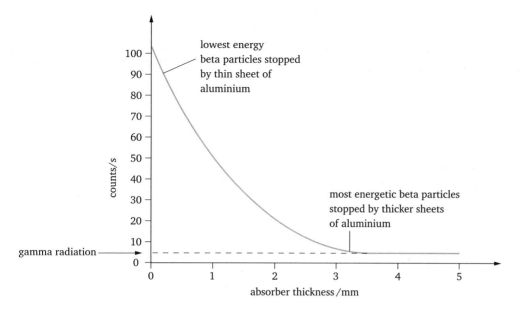

**Figure 2.95**

### Beta particle emitter experiments

- Put the detector a few centimetres from the source and determine the count rate. Do not change these positions.
- Measure the count rate using aluminium absorbers of different thicknesses inserted between the source and the detector.
- Measure the thickness of the absorber using a micrometer.

The count rate will fall as the thickness of the absorber increases, producing a graph similar to Figure 2.95.

The flat portion starts when all the beta particles have been absorbed. This will be when a few millimetres of aluminium have been inserted. The flat portion results from any gamma radiation that is emitted by the source.

### Gamma radiation experiments

This is similar to the beta particle emitter experiment but lead is used as an absorber instead of aluminium. You can do the experiment with aluminium but you will need a lot more of it.

It is necessary to have sufficient distance between the source and the detector to allow insertion of the maximum thickness of lead that is to be used.

A typical graph for this experiment is shown in Figure 2.96.

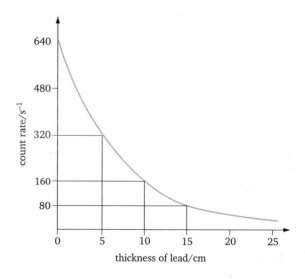

**Figure 2.96**

The graph should show that it always takes the same thickness of lead to halve the count rate. You should determine this value from your graph. The graph shows the count rate to be decreasing *exponentially* with thickness.

## Inverse Square Law for Gamma Radiation

As with all electromagnetic radiation, the intensity varies according to an inverse square law (see page 59). This assumes that there is no absorption by the medium through which it is travelling.

When using gamma rays in air the absorption is low and the count rate $C$ should obey the equation:

$$C = \frac{\text{constant }(k)}{r^2} \quad \text{or} \quad r = \frac{k}{\sqrt{C}}$$

where $r$ is the distance between the source, which is somewhere inside the source container, and the point of detection, which is somewhere inside the Geiger–Müller tube.

In Figure 2.97 only the distance $d$ can be measured. The extra distance $d_0$ between the source and the point of detection is unknown.

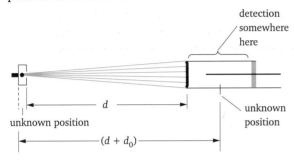

**Figure 2.97**

However, $\dfrac{k}{\sqrt{C}} = d + d_0 (= r)$ so a graph of $\dfrac{1}{\sqrt{C}}$ against $d$ produces a straight-line graph if an inverse square law is obeyed (Figure 2.98).

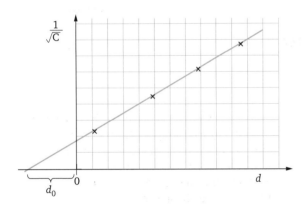

**Figure 2.98**

The extra (unknown) distance $d_0$ is the negative intercept on the $d$-axis.

### Test your understanding

1. (a) What is meant by the range of an alpha particle?
   (b) Why is it necessary to use half thickness instead of range when considering the penetrating power of gamma radiation?
2. A source is thought to emit alpha and gamma radiation but no beta radiation. Describe how you would check whether this was true or not.
3. (a) Explain what is meant by an 'inverse square law'.
   (b) Why would you expect gamma radiation to obey an inverse square law?
   (c) How does the fact that gamma radiation obeys an inverse square law give rise to a precaution necessary when using gamma ray sources?

## Physiological Effects of Radiation

Damage to cells may affect an individual by producing:
- death due to cancer
- radiation sickness (nausea, loss of hair, tiredness)
- damage to genes that are transmitted to offspring and may produce mutations or other harmful effects.

Ions are very reactive and there may be chemical changes in a body cell that interfere with the operation of the cell. A molecule could break apart if a bonding electron is lost and when a large number of cells are destroyed by high doses of radiation the body may not be able to replace them quickly enough. A cell may therefore die or become defective. The defective cell may then divide and produce more defective cells. This rapid production of defective cells is cancer.

## Effect of Radiation on Materials

High radiation levels can affect materials, making them brittle and therefore weaker than usual. Account has to be taken of this in materials used in nuclear reactors or spacecraft that move in high levels of cosmic radiation.

## Some Industrial Uses of Radioactive Sources

### Controlling thickness of sheet metal or paper

Figure 2.99 shows systematically how this is achieved for paper.

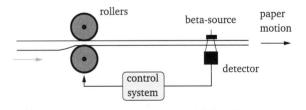

**Figure 2.99**

The source and detector are placed so that the sheet runs between them. If the sheet thickness increases, the count rate falls. This causes an increased pressure on the rollers, which reduces the thickness of the sheet to the desired thickness.

The source used depends on the material. In each case a long half-life is preferable so that the source does not need to be replaced frequently. For paper a beta source is suitable. No alpha particles could get through the paper and the intensity of gamma radiation would not be changed enough when the thickness of paper changes.

Only gamma radiation will work for measuring metal thickness. Alpha particles will be totally absorbed. Beta particles will also be absorbed unless the metal is very thin.

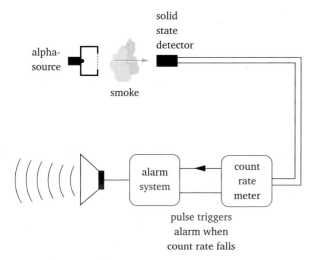

**Figure 2.100**

### Smoke alarms

Figure 2.100 shows a system used for a smoke alarm.

Normally radiation from the long half-life alpha source falls on the detector, producing a steady count rate. Smoke particles are dense enough to cause the count rate to fall. The fall triggers electronic circuitry and an alarm is sounded.

### Monitoring the environment

A short half-life radioactive source is added to a material (e.g. effluent from a sewerage plant or industrial waste). Its presence at various places (e.g. down river from the plant) can be detected and the movement of the effluent monitored.

## Some Medical Uses of Radioactive Sources

### Diagnosis

The body uses the radioactive isotope of an element in the same way as a non-radioactive isotope of the element and therefore a radioactive isotope can be used as a **tracer**. A patient is given an injection or drink containing a low dose of a short-lived radioactive isotope. Its movement is then detected externally as it moves around the body.

For example, radioactive iodine-131 can be used to provide information about the effectiveness of a patient's thyroid gland.

### Therapy

Although radiation can produce cancerous tumours, it can also be used to treat a patient who has a tumour. The risks associated with the treatment have to be balanced against the benefits.

Gamma radiation from cobalt-60 is used to irradiate and kill cancerous cells. The radiation is directed at the tumour from different angles during treatment (see Figure 2.101) so that although the tumour receives a high dose of radiation, the radiation incident on normal cells is as low as possible.

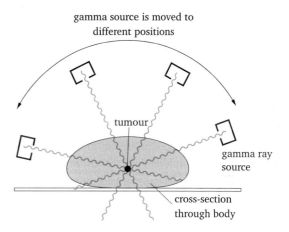

**Figure 2.101**

### Test your understanding

1. A small source emits gamma rays. The count rate is 280 s$^{-1}$ when the Geiger–Müller tube is 10 cm from the source. What would you expect the count rate to be when the tube is:
   (a) 20 cm from the source
   (b) 15 cm from the source?
2. Describe briefly one industrial and one medical use of radioactive materials. State the emission and the half-life of the source that is most appropriate for the applications you have chosen.
3. An isotope of americium decays by the emission of an alpha particle to form neptunium-237. The proton number of neptunium is 93.
   (a) Determine:
      (i) the proton number of the americium isotope
      (ii) the nucleon number of the americium isotope.
   (b) Describe in detail a test that you could carry out to demonstrate that the emission from americium is in fact alpha radiation.

continued ▶

4. The energy of an alpha particle is $8.8 \times 10^{-13}$ J. The mass of a proton or neutron is $1.7 \times 10^{-27}$ kg. Calculate the speed of the alpha particle.

5. (a) Discuss briefly the relative penetrating powers of alpha, beta and gamma radiation.
   (b) Explain what is meant by external and ingested radiation.
   (c) Explain why gamma radiation is the most dangerous external radiation.
   (d) Why in general is ingested radiation more dangerous than external radiation?
   (e) List four important safety precautions that should be taken when conducting experiments with radioactive materials.

# Probing the Nucleus

Nuclei are very small: the largest are approximately $10^{-14}$ m in diameter. Because this is far smaller than the wavelength of light, it is impossible to see into the nucleus of an atom.

The structure of the nucleus can be investigated by inspecting what happens when it is bombarded with high energy particles. In early experiments, the only particles available were alpha particles from the decay of radioactive materials. The first indication about the structure of the atom was found in this way.

## The Existence of the Nucleus

In 1910, Geiger and Marsden, working with Rutherford at the University of Manchester, directed alpha particles (helium nuclei) at a strip of very thin gold foil and determined the directions of the emerging particles (Figure 2.102).

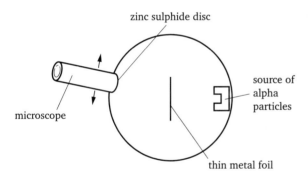

**Figure 2.102**

The deflected particles hit a small disc of zinc sulphide mounted at the focus of a microscope that was pointed at the foil and a tiny flash of light occurred at each hit. The observer had to count the number of flashes occurring in a given time – a difficult and tiring job!

It was found that:
- most of the particles were not deflected
- some of the particles were deflected through very large angles, up to 180° (Figure 2.103).

Knowing that alpha particles were positively charged and had a significant mass, it was concluded that:
- most of an atom is empty space
- the mass of the atom is concentrated in a very small nucleus
- this nucleus is positively charged.

Geiger and Marsden assumed that electrostatic repulsion caused the deflections. The nucleus must be very small for its electric field to be 'concentrated' enough to deflect the alpha particles so much. However, the nucleus also had to be very massive itself since it was not knocked away by the alpha particles.

Since the nucleus is so massive and so small, the electrons surrounding the nucleus must be spread out, leaving most of the atom as empty space.

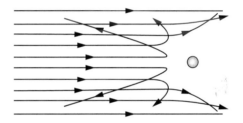

**Figure 2.103**

## Protons in the Nucleus

In 1919, Rutherford observed that protons were expelled from nitrogen nuclei that were bombarded by alpha particles (Figure 2.104).

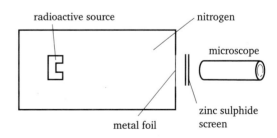

**Figure 2.104**

It was clear that the particles observed hitting the zinc sulphide screen could not be alpha particles as these would be absorbed by the metal foil. The nature of the protons was confirmed by observations in a **cloud chamber**.

## Cloud Chambers

A cloud chamber contains a saturated vapour that is cooled below the temperature at which it would normally condense. The vapour will only condense when there is something to condense on to. In the absence of dust in the chamber, the vapour will condense on gas ions.

When a charged particle goes through the chamber, it ionizes the gas as it goes. The vapour condenses onto the trails of ions, leaving a track that can be photographed and measured (see Figure 2.105).

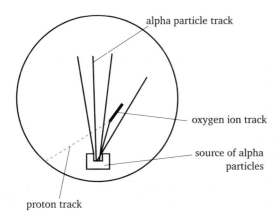

**Figure 2.105**

Tracks in the cloud chamber are characteristic of the particle causing them:
- alpha particle tracks are quite thick because they cause a lot of ionizations per centimetre of their length
- alpha particle tracks are only a few centimetres long because they lose energy, making many ionizing collisions with gas atoms
- alpha particle tracks are straight because they have a large momentum and so are not deflected easily
- beta particle tracks are longer and thinner because they make fewer ionizing collisions per centimetre and so cover a longer distance before they lose all of their energy.

In Figure 2.105, the alpha particle is observed colliding with a nitrogen nucleus and being absorbed by it. The nucleus emits a proton. The proton track is thinner than the characteristic alpha particle track as the proton is a lighter particle.

$$^{14}_{7}N + ^{4}_{2}He \rightarrow ^{17}_{8}O + ^{1}_{1}H$$

## The Discovery of Neutrons

The original concept of the nucleus was that it contained protons and electrons. There were difficulties with this idea but these were resolved with the discovery of the neutron by Chadwick in 1932 (Figure 2.106).

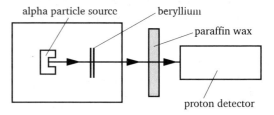

**Figure 2.106**

It was found that the bombardment of beryllium produced a new type of particle that had a very slight ionizing effect. When the new particle hit paraffin wax, it knocked the protons (hydrogen nuclei) out of the wax. These protons could then be easily detected.

## Bubble Chambers

Bubble chambers contain liquids that are superheated. This means that they are heated above their normal boiling temperature. They do not boil until an impurity or disturbance is introduced. The disturbance caused by the passage of a high-energy particle is enough to cause bubbles to form along the length of the path. The bubble track can be photographed and used to identify the nature of the particle.

Figure 2.107 shows a photograph of a bubble chamber in which an antiproton hits and annihilates a proton. Positive and negative particles are produced.

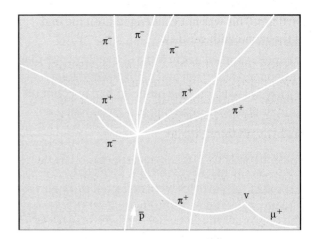

**Figure 2.107**

## Confirmation of Particle Types in Cloud and Bubble Chamber Pictures

If a strong magnetic field is applied to the cloud chamber, the charged particles move in circular arcs (Figure 2.107).

The diagram shows a bubble chamber with a magnetic field directed vertically downwards into the plane of the paper.

Positively charged particles curve anticlockwise and negatively charged particles curve clockwise.

The radius of the curve depends on the speed charge and the mass of the particles:
- the tracks of heavy particles have larger radii
- the tracks of particles with greater charges have smaller radii.

## 'Seeing' the Particles

Leptons such as electrons have no size at all; even atoms are only about $10^{-10}$ m across. Nuclei and baryons such as protons have a diameter of about a hundred thousandth that of an atom. However, charged particles ionize atoms, and ionized atoms have some special properties:
- they act as condensation nuclei in a vapour
- they act as bubble-nuclei in a liquid.

Cloud chambers make use of the first property and bubble chambers make use of the second.

Charged particles moving through the chambers cause droplets/bubbles to form along their tracks.

## Classification of Particles

Fundamental particles cannot be divided into other particles: they have no internal structure. The Greeks considered atoms to be fundamental particles. This was not challenged until the end of the nineteenth century.

It is now known that both protons and neutrons have internal structures: they are composed of **quarks**. At the moment, all the evidence suggests that quarks are fundamental, i.e. they are not made of anything smaller.

There are only six types of quark and most of the material with which we are familiar is made of only two of these.

The only other fundamental particles seem to be **leptons**, of which there are also six types.

For each type of matter particle (leptons and quarks) there is an antiparticle that is similar to the original particle but has the opposite charge. When a particle and its own antiparticle meet, they annihilate each other and produce a burst of energy.

The antiparticle of the electron is the positron, which has the same mass as an electron but has a positive charge. Similarly, an antiproton is negatively charged but has the same mass as a proton.

All matter seems to be made up of combinations of these particles and antiparticles.

## Particle Properties

Some of the properties that distinguish the different particles are **charge**, **mass** and **lepton** or **baryon number**.

## Leptons

Leptons have no linear dimensions, they just occupy a point in space.

There are three groups of leptons:
- electrons, e
- muons, μ
- tau particles, τ

Each group has a neutrino (and all the particles have their antiparticles), making 12 leptons in all:

$e^-$ = electron
$e^+$ = positron ($e^-$ antiparticle)
$\mu^-$ = muon
$\mu^+$ = antimuon
$\tau^-$ = tau
$\tau^+$ = antitau
$\nu_e$ = (electron)neutrino
$\bar{\nu}_e$ = antineutrino
$\nu_\mu$ = mu-neutrino
$\bar{\nu}_\mu$ = mu-antineutrino
$\nu_\tau$ = tau-neutrino
$\bar{\nu}_\tau$ = tau-antineutrino

In Table 2.7 the following symbols are used:

$m$ = relative mass
   (electron mass = 1)
$Q$ = relative charge
   (electron charge = $-1$)
$l$ = Lepton number

## Table 2.7

| Particle | L | m | Q |
|---|---|---|---|
| $e^-$ | +1 | 1 | −1 |
| $\nu_e$ | +1 | 0 | 0 |
| $e^+$ | −1 | 1 | +1 |
| $\bar{\nu}_e$ | −1 | 0 | 0 |
| $\mu^-$ | +1 | 200 | −1 |
| $\nu_\mu$ | +1 | 0 | 0 |
| $\mu^+$ | −1 | 200 | +1 |
| $\bar{\nu}_\mu$ | −1 | 0 | 0 |
| $\tau^-$ | +1 | 3500 | −1 |
| $\nu_\tau$ | +1 | 0 | 0 |
| $\tau^+$ | −1 | 3500 | +1 |
| $\bar{\nu}_\tau$ | −1 | 0 | 0 |

Neutrinos have no charge and infinitesimal mass. They are extremely difficult to detect.

## Muons and Tau Leptons

Electrons are bound to nuclei by electrostatic attraction. Muons and tau leptons are not bound to anything and are common in cosmic radiation.

Muons and tau leptons each carry a charge of −1. The mass of a muon is 200 times that of an electron and the mass of a tau lepton is 3500 times the mass of the electron (twice the mass of a proton).

A muon decays in about 2 μs into an electron, a mu-neutrino and an electron antineutrino.

Apart from their different masses, muons and tau leptons have all the properties of electrons.

## Mesons

Mesons are particles with masses intermediate between those of an electron and a proton. They also play a part in nuclear bonding but their characteristics are akin to those of a proton.

## Hadrons

Hadrons are particles that can be split and have finite (but very small) linear dimensions. Baryons and mesons are subgroups of the hadron family.

**Baryons** are 'heavy' hadrons and include the proton, the neutron and their antiparticles.

**Mesons** are less heavy hadrons. They interact with protons and neutrons inside the nucleus to bind the nucleus together.

## Quarks

The existence of quarks was suggested in the 1960s to account for the properties of hadrons. Quarks, like leptons, come in three pairs and each quark has an antiquark. Also like leptons, they are point bodies and appear to be structureless. Experimental work in the 1960s and 1970s gave some confirmation of their existence.

### Up-quarks and down-quarks

This are the only pair of quarks with which we are concerned. They have fractional charges:

up-quark $\quad$ $\frac{2}{3}e$ and positive

down-quark $\quad$ $\frac{1}{3}e$ and negative

Each baryon consists of three quarks. Hence, a proton consists of two up-quarks and one down-quark (Figure 2.108).

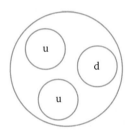

**Figure 2.108**

Each meson consists of a quark plus an anti-quark. For example, an up-quark and an antidown-quark results in a meson called a pion $\pi^+$ (charge +1).

When quarks combine they always do so in such a way that the total charge is a whole number.

## Baryon Number

Just as with leptons, each baryon has a number associated with it and this number is the sum of the baryon numbers of the quarks comprising it:

the baryon number of up- and down-quarks is $+\frac{1}{3}$

the baryon number of each of antiquark is $-\frac{1}{3}$.

Since a meson is a quark and an antiquark it follows that its baryon number is 0.

## Stability

The proton and the antiproton are the only hadrons that are stable. Among the leptons, electrons and neutrinos are stable.

All the other particles have short lifetimes ranging from a few seconds down to $10^{-12}$ s. In other words, they 'decay'.

## The rules governing particle decay or interaction

These rules are essentially conservation laws. In addition to conservation of energy and charge, there is conservation of lepton number and baryon number.

## The decay of a neutron

A neutron, n, and a proton, p, can be expressed in terms of up-quarks, u, and down-quarks, d:

$$n = u + d + d$$
$$p = u + u + d$$

When a neutron decays into a proton, a d-quark must become a u-quark.

If the d-quark decays only into a u-quark and an electron:
- charge will be conserved: $-\frac{1}{3} = +\frac{2}{3} + -1$
- the baryon number will be conserved: $\frac{1}{3} = \frac{1}{3} + 0$
- the lepton number will **not** be conserved: $0 = 0 + 1$, **error!**

There must therefore be another product. If there is only one further product it must have charge 0 and baryon number 0, i.e. it is a lepton without charge. Using Table 2.7, we can see that it must be a kind of neutrino and, as its lepton number must be $-1$ to balance the $+1$ of the electron, it must be an antineutrino. Logic suggests an electron antineutrino, giving:

$$d \rightarrow u + e^- + \overline{\nu}_e$$

We could set the **property equations** below the **decay statement**, as follows:

| | d | $\rightarrow$ | u | + | $e^-$ | + | $\overline{\nu}_e$ |
|---|---|---|---|---|---|---|---|
| Q | $(-\frac{1}{3})$ | = | $(+\frac{2}{3})$ | + | $(-1)$ | + | 0 |
| $B_n$ | $\frac{1}{3}$ | = | $\frac{1}{3}$ | + | 0 | + | 0 |
| $L_n$ | 0 | = | 0 | + | $(+1)$ | + | $(-1)$ |

### Test your understanding

1 Using the fact that up-quarks have a charge of $+\frac{2}{3}$ and down quarks have a charge of $-\frac{1}{3}$, draw a diagram to show the quark structure of a neutron.
2 Use conservation rules to show that a muon could decay into an electron, a muon-neutrino and an electron-antineutrino but not into an electron, a muon-neutrino and an electron-neutrino.
A lepton always decays into its own neutrino and a lighter lepton.
3 State and explain what the charge is on any meson.

# Nature of Information

Information takes a variety of forms. We associate information most commonly with the transmission of reports and ideas in words and images. This type of information is transmitted using newspapers, books, radio, television, CDs and the internet.

However, in their work scientists, engineers and doctors need other types of information. They depend on measurements and observations to develop theories, design new machines, monitor processes and diagnose disease.

These may be any measurements that you are used to making in a laboratory, such as:
- temperature
- position
- speed
- acceleration
- flow rate
- light intensity
- magnetic field strength
- activity of a radioactive source.

Often the measurements may be made directly but sometimes it is advantageous to use data-capture techniques.

## Advantages of Data Capture

Scientists, engineers and doctors frequently want to know how a measurement changes with time and to take measurements in places that are not readily accessible.

For time-dependent measurements a sensor is connected to data-logging devices that take readings at predetermined intervals.

This is advantageous because:
- when events take a short time humans cannot take readings often enough
- when events take a long time humans may not wish to waste time waiting to take readings at suitable intervals
- it may not be known when an event is likely to occur, so the person wanting to take the measurement may not be present.

Remote monitoring of data is necessary for measurements in hostile or inaccessible places such as outer space, inside nuclear reactors, inside the body or inside the Earth.

For example:
- Nuclear reactor engineers need to know such things as temperatures, levels of radioactivity and flow rates of coolants.
- Doctors can provide patients with instruments that monitor the heart. When a patient detects an irregular heartbeat this can be recorded in a data-capture device. The recording can be taken to the doctor for checking.
- Doctors can monitor blood-flow rates using ultrasound sensors.
- Geophysicists and astronomers can measure temperatures, radioactivity and magnetic fields strengths, etc.

## Analogue Instruments

These are instruments that allow readings of any value to be taken. They are continuous and their only limitation is the user's ability to tell when a reading has changed. This depends on the sensitivity of the instrument.

Instruments that give continuous scale readings are called **analogue instruments**. Examples include:
- mercury in glass thermometers
- moving coil ammeters
- rulers.

## Computer Monitoring and Remote Sensing

The following processes enable readings to be taken and fed into a computer to be read remotely (i.e. a long way from the place where the measurement is taken).

The instrument reading is converted into a corresponding potential difference. The voltage is transmitted directly to the monitoring equipment.

The potential difference is measured directly on a voltmeter calibrated with a suitable scale. It is then converted into a digital form that provides a **digital readout** (using a seven-segment display) or stored in a data-logging device. The digital information may be saved for inspection later or used directly by a computer in real time.

## Sensors

Potential difference can be measured directly and current can be measured by measuring the voltage across a fixed known resistor.

For other measurements, a suitable sensor is needed. The sensor must be capable of being included in a circuit that will produce a corresponding voltage. Such a system is shown schematically in Figure 2.109.

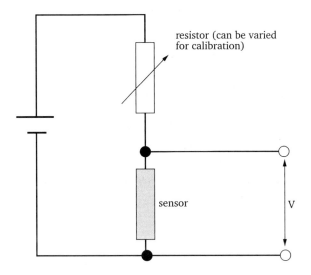

**Figure 2.109**

For example:
- position can be measured with a linear or rotational potentiometer
- temperature can be measured with a thermistor
- light intensity can be measured with a light-dependent resistor
- magnetic fields can be measured with a device called a Hall probe.

Sensors can be used to measure flow rates of liquids or gases, radioactivity levels, etc.

## Current Measurement

The principle of conversion to a potential difference is fairly easy to see in the system to measure current shown in Figure 2.110.

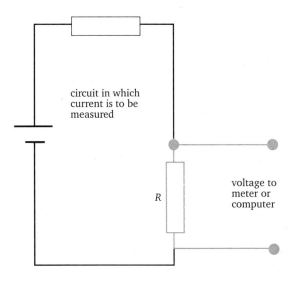

**Figure 2.110**

The sensor is the fixed known resistor of resistance $R$ (e.g. 10 $\Omega$).

The potential difference $V$ measured across the resistor is given by:

$$V = IR$$

$V$ is directly proportional to $R$, so all that is needed is to put a scale on the voltmeter (or calibrate the computer) to read current instead of voltage.

If $R$ is 10 $\Omega$ then 1 V corresponds to a current of 0.1 A (100 mA).

Other sensors, such as the position sensor, may use circuits that produce a constant current through them.

In a **position sensor**:
- the sensor may be a potentiometer used as a variable resistor
- as the slider moves the resistance changes
- the potential difference $V$ is directly proportional to the resistance of the sensor
- the potential difference depends on the slider position
- calibration is necessary to define the change in position for a given change in voltage.

Alternatively, the potential divider principle can be used to produce a potential difference as shown in Figure 2.111.

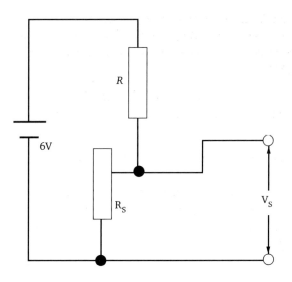

**Figure 2.111**

The potential difference across the sensor is given by:

$$V_s = \frac{R_s}{R_s + R} \times V$$

where $R$ is the resistance of the series resistor and $V$ is the supply voltage.

If the series resistance is large, the voltage is approximately proportional to $R_s$.

EXAMPLE ONE: Figure 2.112 shows how the resistance of a temperature sensor varies with temperature.

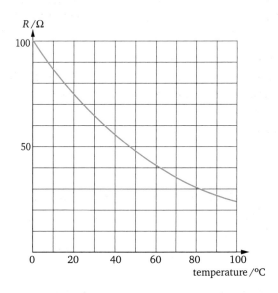

**Figure 2.112**

The sensor (a thermistor) is placed in the circuit shown in Figure 2.113.

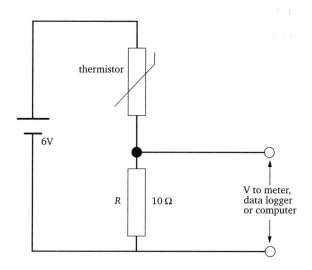

**Figure 2.113**

When the temperature is 0°C, the resistance of the sensor is 100 Ω. The potential difference sent to the data logger is therefore 0.55 V.

When the temperature is 100°C, the resistance of the sensor is 30 Ω and the potential difference is 1.50 V.

By varying the value of the resistor $R$ the voltage corresponding to 100°C could be made equal to 1.0 V.

**Test your understanding**

1. (a) Calculate the value of the resistor $R$ for the potential difference at 100°C to equal 1.0 V.
   (b) Calculate the resistance at 0°C when the resistor has this value.

## Digital Data

Digital readouts and computers need information in the form of 0s and 1s. (These are actually two voltage levels, 0 and 5 V.) Binary numbers are used for this purpose.

Table 2.8 shows how numbers are represented in binary form up to 10.

**Table 2.8**

| Decimal | Binary |
|---------|--------|
| 0 | 0000 |
| 1 | 0001 |
| 2 | 0010 |
| 3 | 0011 |
| 4 | 0100 |
| 5 | 0101 |
| 6 | 0110 |
| 7 | 0111 |
| 8 | 1000 |
| 9 | 1001 |
| 10 | 1010 |

An analogue-to-digital converter is used to convert a voltage level into a digital form.

Figure 2.114 shows how different voltage levels are represented. The graph assumes that there are only 10 possible voltage levels that can be identified. In practice there are 127.

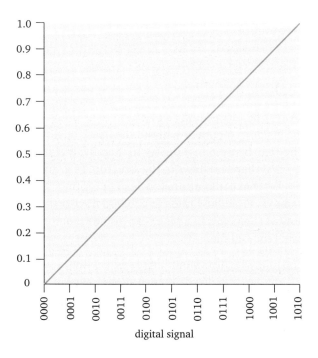

**Figure 2.114**

Assuming that the maximum voltage to be measured is 1 V, then:
- any voltage between 0 and 0.1 V will be represented as 0000
- when the voltage reaches 0.1 V the binary count triggers to become 0001
- any voltage between 0.1 and 0.2 V will be represented as 0001
- when the voltage reaches 1 V the binary count triggers to become 0010 and so on.

There is therefore an uncertainty in the last digit since a voltage reading of 0001, for example, means that the true voltage is anywhere between 0.1 and 0.2 V.

If the maximum voltage in this example were 2 V, each increment would represent 0.2 V so the resolution would be worse.

In practice digital numbers up to 127 (binary 11111111) are used. More voltage levels can be detected in a given range, which improves the precision of the digital data, but you must remember this limitation when using data-capture techniques.

## Sampling Data

A computer stores the binary number in its memory with a record of the time at which the data was recorded. The only limitation on the amount of data stored is the amount of computer memory available.

An experiment may last from a fraction of a second to many days or even weeks. You decide the sampling rate or the duration of the experiment.

The data logger or computer takes readings at suitable time intervals. The logger uses its own internal clock to record the time at which the data is sampled. The more frequently data is sampled the more information you have, but sampling too often in a long experiment may use more memory than is justified by the experiment.

When recalled from memory the digitally stored data is converted back to decimal form using a binary to decimal converter. The data can drive seven-segment displays or be presented in a table or as a graph on a monitor.

## Transmitting Data

Data can be transmitted as a stream of 0s and 1s along copper cable or along optical fibres or by radio waves.

### Copper cable

In computers the 0s and 1s can be sent along cables connected in parallel or as a string along a single cable.

In a parallel cable each bit (binary digit) of the number arrives at the same time. In a serial cable the numbers are sent as a stream one after the other. Figure 2.115 shows successive binary numbers on a cable at a given time on a parallel and on a serial cable.

A **byte** represents one piece of information. In data capture a byte will represent a number.

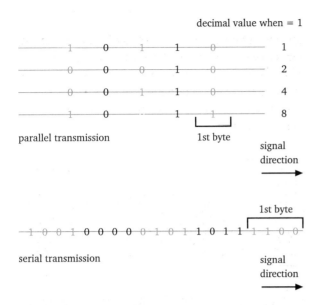

**Figure 2.115**

A major problem with cables is that the currents in the wires cause heating and therefore a loss of energy ($I^2R$ losses). Also, the voltage level falls. Electronic repeaters have to be installed frequently to restore the signal strength as it travels along. This is illustrated in the following example.

EXAMPLE TWO: A thin copper cable has a resistance of 1.5 Ω m$^{-1}$. The voltage at the sensor is 5.0 V. The current carried by the cable transmitting a digital signal is 12 mA.

(a) Calculate the voltage drop per metre length of the wire.
(b) When would a repeater be needed if the signal voltage must not drop below 3.5 V?
(c) How much power is used in heating the wires between the repeaters?

(a) Voltage drop per m = $1.5 \times 12 \times 10^{-3}$ V
    = $18 \times 10^{-3}$ V
(b) Voltage drop allowed = $5.0 - 3.5 = 1.5$ V
    Length of cable for this drop = $1.5/18 \times 10^{-3}$
    = 83 m
(c) Resistance = $1.5 \times 83 = 124$ Ω
    Power used = $0.012^2 \times 124 = 0.018$ W

### Fibre optic cable

When using fibre optic cable the digital signals are sent serially along the cable as a series of light pulses: light ON = 1 and light OFF = 0. This is achieved by switching a light-emitting diode on and off as required.

Fibre optic cable can be made very pure so that little light is absorbed in the transmission of the light pulse. The signal can therefore travel further before it needs to be boosted by a repeater. The signal is secure since the only way of extracting information illegally would be by cutting the cable (which would be noticed by the user).

Materials (silicon from sand) for fibre optic cables are cheaper than for copper cables but this has to be offset against more complex manufacturing techniques.

## Thermal Energy and Miniaturization

Although the currents in microcircuits are small, the components and wires used are finer and there are more of them in a small space. There can therefore be a lot of thermal energy generated per unit volume in a microchip. Heating damages the circuitry in the microchip so cooling is essential.

Most of the noise generated by a computer is the cooling fan keeping the processor cool by blowing air across metal fins attached to the processor to aid the energy transfer.

Trying to pack more into a small space results in more thermal energy and the need for more efficient cooling.

## Information using X-rays and Ultrasound – Comparison of Techniques

X-rays and ultrasound can be used to examine the inside of a patient without the need for surgery. They can be used in industry to examine metals to check if there are flaws in the structure (e.g. small cracks) that could ultimately lead to serious damage. Fluid flow rates (e.g. blood flow or the flow of chemicals in a pipe) can also be monitored using the Doppler effect.

The following lists summarize some of the advantages and disadvantages of each type of wave.

X-rays:
- are highly penetrating, so pass through a patient's body
- are absorbed more by high density and greater thickness of material
- are usually used to provide information in the form of a photographic image using a transmitted radiation photographic plate
- can provide an image of the inside of a suitcase in airport security
- are ionizing, so there are limitations on how long and how often they can be used without harm to a patient
- have short wavelengths ($10^{-10}$ m) so provide better detail of small objects than ultrasound
- require careful precautions by the user (a radiologist) to avoid overexposure to the radiation over time.

Ultrasound:
- is reflected from organs, etc.
- is less penetrating than X-rays so diagnosis is difficult with obese patients where the organ to be investigated is deep inside the body
- is non-ionizing so is safe for patient and user to use over extended periods of time to obtain more information
- is produced by a source that can be moved freely around the patient's body surface to obtain more and better information
- is able to produce real-time video images on a monitor of the function of organs such as the heart
- can be used to investigate the motion of fluid such as blood in the body using the Doppler effect.

**Test your understanding**

1. The following sequence of data, sampled at a rate of 1 byte per ms, was received along a simple parallel cable. The signal represented by 1010 was 5 mV. The signal to the right was received first. Sketch on graph paper the reconstituted analogue signal.

| 0 | 1 | 1 | 0 | 0 | 0 | 1 | 0 | 0 | 0 | 0 | 1 | 1 |
| 1 | 0 | 0 | 1 | 1 | 1 | 0 | 1 | 0 | 0 | 1 | 0 | 0 |
| 1 | 0 | 0 | 0 | 0 | 0 | 0 | 1 | 1 | 1 | 1 | 0 | 1 |
| 0 | 1 | 0 | 1 | 0 | 1 | 0 | 1 | 1 | 0 | 0 | 1 | 0 |

2. The resistance per metre of a copper cable is 25 mΩ. The voltage drop per metre for this cable when used to transmit a signal is 0.15 mV.
   (a) Calculate the voltage drop across 250 m of the cable.
   (b) Calculate the current in the cable.
   (c) Calculate the power lost in transmission using 250 m of this cable.
   (d) A signal is sent that has an initial amplitude of 3.20 V. Determine the amplitude after travelling 250 m.
   (e) The signal must not fall below 2.5 V. What is the maximum length of cable that could be used without a repeater?

3. (a) Explain how you would set up a rotational wire-wound potentiometer as a sensor to convert angular measure into a voltage.
   (b) Suggest the factors that will affect the smallest change in angle that could be measured.

4. Write down two significant advantages and two disadvantages for transmission of information using:
   (a) copper cable
   (b) fibre optic cable.

## Communication Systems

The need to pass messages over a distance has been necessary since humans started communicating. What has changed over the years is the speed, range and amount of information that can be transmitted in a given time. Figure 2.116 shows how these factors have changed over time.

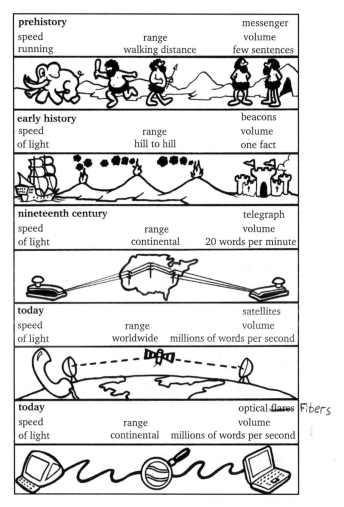

**Figure 2.116**

Each of the ways of communicating shown in Figure 2.116 is a communication system. Figure 2.117 is a block diagram showing the components of a simple communication system.

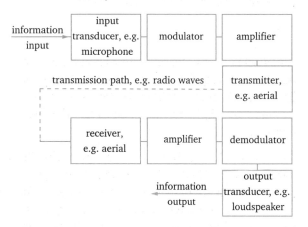

**Figure 2.117**

- Information from the originator of the information (in this case audio) is changed into voltage variations by a transducer (a microphone).
- In the modulator the information is superimposed on a carrier frequency as a variation in amplitude or frequency.
- The voltage variations produced by the microphone are amplified and are fed to the transmitter.
- The information is transmitted as a radio wave directly through space to a receiver or via a satellite. Alternatively, fibre optic cable or metal cable may be used. The speed of transmission is $3 \times 10^8$ m s$^{-1}$ through space but it may be $\frac{2}{3}$ of this in fibre optic cable.
- At the receiver the electromagnetic waves are converted back into voltage variations.
- A demodulator retrieves the information from the carrier.
- The information signal is amplified and used to drive an output transducer (a loudspeaker).

For some purposes it is not necessary to reproduce the information precisely as long as the 'message' can be understood. In other instances loss of detail in the transmitted information is not acceptable. To understand the different types of communication systems in use it is necessary to appreciate the nature of the information that is to be transmitted.

## Audio Information and Bandwidth

### Range of hearing

The human ear is sensitive to frequencies in the range of about 15 Hz to 20 000 kHz. The ear is most sensitive at about 3000 Hz. This is the resonant frequency of the cavity in the ear that leads to the ear-drum.

The range of hearing depends on the observer and decreases with age. It may be reduced due to damage produced by exposure over a long period to loud sounds such as pneumatic drills or very loud music.

### Simple sound information: frequency spectrum

Sound information may be very simple. For example, the sinusoidal variation of the displacement of air against time, shown in Figure 2.118a, represents a note that is a single frequency.

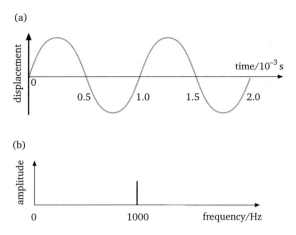

**Figure 2.118**

The sound can be represented by a **frequency spectrum**. This shows the frequencies that are present, plotted on the *x*-axis, and their relative amplitudes, shown by the height of the vertical line. Figure 2.118b shows the frequency spectrum for the pure note in Figure 2.118a.

Although a single note consisting of a single frequency is not in itself very useful information (the sound is either there or not there), this technique has been used in the transmission of information using Morse code, where the length of the transmission of the frequency produces a 'dot' or a 'dash'. Digital information could also be transmitted by a single frequency, for example no signal could represent 0 and a received signal could represent 1. In practice, two tones are used to represent 0 and 1.

## Analogue Information

The displacement of a sound wave varies continually with time. When transmitting analogue information a voltage is transmitted that is proportional to the displacement. The transmitted voltage variation is a replica of the displacement time variation. This voltage variation is converted back to the original signal at the receiver.

## More Complex Sounds: Speech Transmission

Figure 2.119 shows the waveforms of displacement against time for a man making the vowel sound 'i'. The corresponding frequency spectrum is also shown. Figure 2.120 shows the corresponding diagrams for a woman.

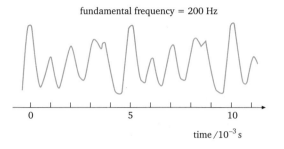

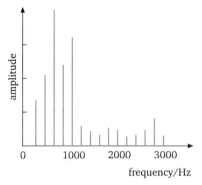

**Figure 2.119**

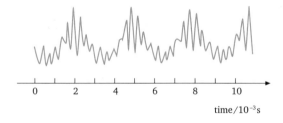

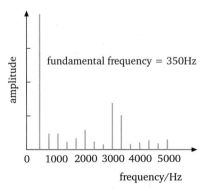

**Figure 2.120**

The male fundamental frequency is lower than that of the female, and neither is a simple sinusoidal wave.

Speech consists of a succession of complex waveforms, such as those in Figures 2.119 and 2.120. The frequency spectrum of the transmitted information is continually changing but the frequencies always lie in the same range.

Different musical instruments playing the same note show similarly complex waves. However, in music the higher frequencies in the frequency spectrum have larger amplitudes and are therefore more important for accurate transmission of the sound.

## Base Bandwidth

The base bandwidth of a communication channel:
- is the range of frequencies present in the original information that is transmitted
- is the range of frequencies that is transmitted for effective transmission
- affects the quality of the sound that is received.

For high quality transmission of music, audible frequencies from 15 Hz to 15 kHz have to be transmitted. High quality also requires elimination of **noise**, which is any unwanted signal. Frequency modulation (FM) is used for such transmissions.

Although all frequencies to which the ear is sensitive are present in speech, most of the frequencies are in the range below about 3.4 kHz. Adequate speech is transmitted in a base bandwidth of 300 to 3000 Hz. This range is used in telephone systems.

For reasonable quality music and good quality speech the base bandwidth range has to be 50 to 4500 Hz. This is the range used when transmitting signals by medium and long waveband radio. Amplitude modulated (AM) transmissions are used in this case.

### Task

Using a radio, determine the frequency between one station and the next in each of the wavebands.

Deduce from your observations the base bandwidth of the information being transmitted.

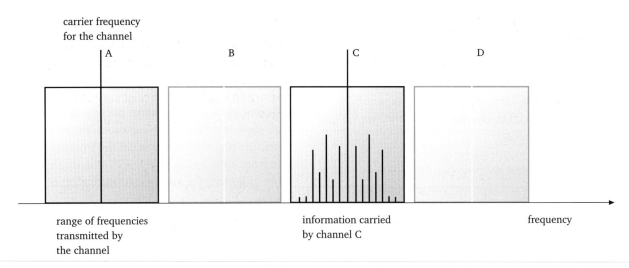

**Figure 2.121**

## Modulation

If base bandwidth signals were transmitted directly using radio waves, only one signal could be sent at a time. The wavelength would also be very long, requiring impossibly long aerials for efficient transmission.

These difficulties are overcome by giving each user a unique frequency to carry the information (**the carrier frequency**). This is the frequency to which you tune to receive the station or channel you want when using radio or television.

This information can be transmitted by modifying:
- the amplitude of the carrier wave (AM)
- the frequency of the carrier wave (FM).

This process is called **modulation**. The information is retrieved at the receiver by a process called **demodulation**. In AM this is achieved easily by the use of a diode.

Figure 2.121 shows how the total range of frequencies available in a given waveband is split into chunks of size equal to the width of a channel. The frequency width of the channel depends on the nature of the information.

## Channel Bandwidth for Audio Information

The technology used in the transmission of information requires a wider bandwidth than the base bandwidth. The transmitted bandwidth is approximately twice the highest frequency in the base bandwidth. Communication channels and radio stations use about
- 6 kHz (telephone)
- 9 kHz (AM radio)
- 200 kHz (FM radio)

The carrier frequency is the centre frequency of the channel.

The waveband for very high frequency (VHF) radio is 88 to 108 MHz. This is used for FM transmission. The total available frequency range is 20 MHz. The maximum number of channels available is $\frac{20 \text{ MHz}}{100 \text{ kHz}}$.

This means that 200 FM channels could be transmitted in this band.

## Video Information

To transmit black and white or colour video (television) information with sound a much wider base bandwidth of about 8 MHz is needed. It follows that far fewer video channels than audio channels can be transmitted in the same range of frequencies.

## Transmitting more Channels

Using higher frequency ranges to carry the information allows more channels to be transmitted. The range available for ultra high frequency (UHF) transmission, used for analogue television transmission, is 300 to 3000 MHz, a total of 2700 MHz. This range could carry $\frac{2700}{28}$ channels, which is approximately 340 colour television channels:

$$\text{the number of channels} = \frac{\text{total bandwidth available}}{\text{bandwidth needed for each channel}}$$

Using optical frequency ranges and transmitting information down optical fibres increases the available frequency range. The range of visible light is about ($5 \times 10^{14}$ Hz).

Theoretically $\dfrac{(5\times 10^{14})}{(8\times 10^{6})} = 6.3\times 10^{7}$ analogue television channels could be transmitted down one optical fibre. Fewer channels are available in practice.

> **Test your understanding**
>
> 1. Radio 4 is transmitted using FM at 92.8 MHz. Calculate the wavelength of the radio wave used for this transmission.
> 2. A radio station is transmitted using AM with a wavelength of 909 m. Determine the range of frequencies transmitted.
> 3. Consider what makes up a colour television image and write down the information that you would need to transmit so that it could be reproduced.
> 4. The first transatlantic cable was laid from Scotland to Newfoundland in 1956. It carried 36 voice channels. Estimate the total range of frequencies carried by the cable.

## Digitizing Information

Analogue information can be converted into digital form for transmission by sampling the amplitude of the waveform at regular intervals.

As analogue signals weaken they become difficult to detect above unwanted electromagnetic radiation (called **noise**). Digital transmission has the advantage that although the signal strength falls, it can be processed to produce a signal as clear as that originally transmitted. The only requirement is that the 1s are detectable above the noise level. This is illustrated in Figure 2.122.

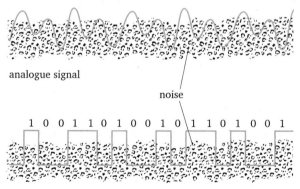

**Figure 2.122**

## Sampling Rates

If the sample rate is too low, variations that occur in the interval between sampling are lost. In Figure 2.123 the fall in the signal at A has been missed. This would result in a loss of quality during transmission.

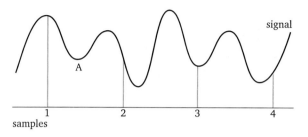

**Figure 2.123**

Doubling the sampling rate, as shown in Figure 2.124, would identify this change and improve the quality of the sound.

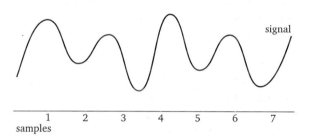

**Figure 2.124**

The minimum sampling frequency is:
- twice the frequency of the highest frequency to be transmitted
- twice the frequency of the maximum frequency in the base bandwidth of the audio or video information
- about 8 kHz for speech
- 40 kHz for good quality music.

Each time the wave is sampled, a train of eight bits of information has to be transmitted. One train of eight bits is called a **byte**.

A **bit** is a 'binary digit', i.e. a 0 or a 1.

One bit is used to synchronize timing between the transmitter and the receiver. The level of the signal strength at the time the sample is taken is in the other seven bits.

## Converting Analogue Voltages to Digital Signals

Figure 2.125 shows a simplified view of the relationship between the analogue waveform and the transmitted digital signal. The analogue waveform could be a voice signal or the output of a sensor.

The signal value at the time of sampling is converted into the binary form using an analogue-to-digital converter.

To simplify the diagram:
- only four bits are sent for each sample
- the maximum of the transmitted signal in the example is represented by decimal 10 (binary 1010)
- the timing bit has been omitted.

The digital bits are sent between one sample and the next. If you dial a fax number instead of a telephone number, the sound you hear is produced by the frequencies of the bits transmitted by the fax machine.

In practice the maximum signal voltage corresponds to binary 1111111 or decimal 127, which leads to a better representation of the signal than in Figure 2.125.

The wave displacement is sampled at 8 kHz. The minimum number of bits per second required in a digital speech transmission is therefore approximately (8 bits × 8 kHz = 64 000 bits per second (64 kbit s$^{-1}$).

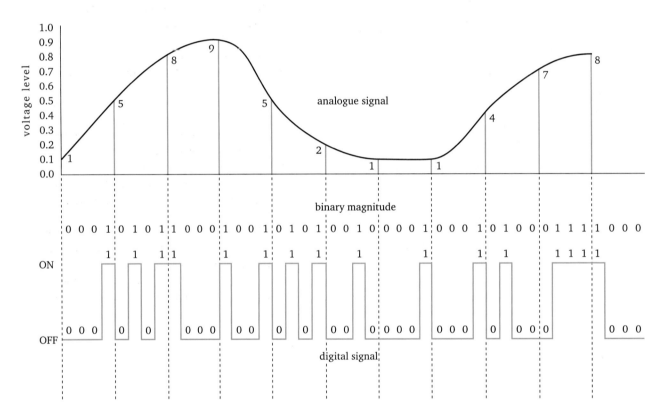

**Figure 2.125**

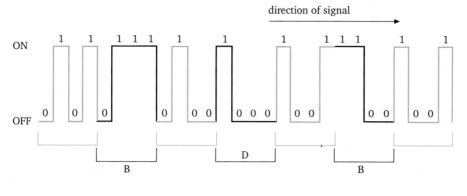

**Figure 2.126**

## Time-division Multiplexing

Bits can be sent at much higher rates than the 64 000 bit s$^{-1}$ required for an audio signal such as in telephones. A coaxial cable can carry 140 Mbit s$^{-1}$. This means that a single coaxial cable can carry

$$\frac{(140 \times 10^6)}{(64 \times 10^3)} = 2200 \text{ telephone conversations.}$$

The cable may be shared by 2200 users, with each user's signal being sampled in turn. Each signal only uses the cable for $\frac{1}{2200}$ of the time, hence the term 'time division'.

Figure 2.126 shows the digital waveform when four digital signals, from users A, B, C and D, are sent down a single wire or optical fibre. (Again for simplicity a system with only four bits per sample is shown.)

### Test your understanding

1. Determine the number of bits per second that need to be transmitted for:
   (a) good quality music
   (b) video information.
2. A transmission system can handle bit rates of 2.4 Gbit s$^{-1}$.
   (a) How many telephone channels could this system handle at any one time?
   (b) How many television channels could be transmitted at a time?
3. A system samples a waveform at intervals of 1 ms. The maximum signal transmitted is 5 V, which is represented by 1010. Each sample is transmitted by four bits of data.
   Write down the sequence of bits that would be transmitted during a 10 ms time interval when transmitting a square waveform that switches between 0.5 and 4.0 V. The sequence should start at the instant when the signal switches to 4.0 V.

# Answers

## Unit 1

### Page 18

1. (a) 5.4 N at 22° to the horizontal.
   (b) 19 N at 18° to the horizontal.
   (c) 270 N at 24° to the horizontal.
   (d) 50 N at 71° to the horizontal.

2. As for question 1.

### Page 19

1. 243 m s$^{-1}$ on a bearing of 008.3°.

2. 12.5 knots on a bearing of 196°.

3. (a) Resultant horizontal component = 30 N to the right; resultant force = 34 N at 27° to the horizontal.
   (b) Resultant horizontal component = 25 N to the left; resultant force = 41 N at 52° to the horizontal.
   (c) Resultant horizontal component = 290 N to the right; resultant force = 490 N at 53° to the horizontal.

4. (a) $F_H$ = 8.5 N; $F_V$ = 8.5 N
   (b) $F_H$ = 340 N; $F_V$ = 120 N
   (c) $F_H$ = 0.20 N; $F_V$ = 0.14 N

### Page 20

1. (a) 27 m s$^{-1}$    (b) 13 m s$^{-1}$

2. (a) A = 3.8 m s$^{-1}$, B = 2.0 m s$^{-1}$
   (b) A, by 57 s
   (c) A = 3.2 m s$^{-1}$, B = 3.4 m s$^{-1}$
   (d) 210 m

3. (a) $W_{parallel}$ = 6.0 N; $W_{perpendicular}$ = 10 N
   (b) $W_{parallel}$ = 260 N; $W_{perpendicular}$ = 20 N

4. (a) 83 N m
   (b) 72 N m

### Page 21

1. 2.5 m

2. (a) See page 20.
   (b) $T_A$ = 113 N; $T_B$ = 132 N

3. (a) 8.3 m s$^{-2}$
   (b) Resultant force is weight–viscous resistance. Viscous resistance increases as speed increases therefore resultant force decreases as speed increases and acceleration decreases since acceleration is proportional to force.
   (c) When speed = $v$, viscous resistance is equal and opposite to weight, resultant force is zero and acceleration is zero so speed is constant.
   (d) 7.0 × 10$^{-3}$ m

4. (a) (i) Acceleration = 0.75 m s$^{-2}$; deceleration = 0.60 m s$^{-2}$
       (ii) 2300 m
       (iii) 60 kN
       (iv) 48 kN
   (b) (i) 41 kN
       (ii) 0.49 m s$^{-2}$

### Page 24

1. (a) 4.0 m s$^{-1}$    (b) 2.0 m s$^{-1}$
2. (a) 3.3 m s$^{-2}$    (b) 1.4 m s$^{-2}$
3. (a) 60 m              (b) 220 m
4. (a) 65 m              (b) 1.7 m s$^{-2}$

### Page 26

1. acceleration = 2.5 m s$^{-2}$; distance = 80 m

2. (a) 5.9 m s$^{-2}$    (b) 6.7 m
   (c) 93.3 m            (d) 12.0 s

3. (a) 25 m              (b) 4.5 s

### Page 27

1. (a) 1.2 s
   (b) 2.5 s
   (c) 7.3 m
   (d) Horizontal distance travelled = 37 m.

2. Maximum height = 17 m
   Horizontal distance = 115 m

3. (a) 2.1 m    (b) 88.7° to the horizontal

## Page 28

1. (a) 5.9 N  (b) 74 N m$^{-1}$
   (c) 0.24 J

2. (a) 170 N m$^{-1}$  (b) 17 N

## Page 31

1. See question 5.

2. See page 30.

3. (a) Gravitational potential energy to kinetic energy
      Gravitational potential energy to internal energy as a result of work done against friction
      Gravitational potential energy of skier to kinetic energy of air
   (b) Gradient of graph at t = 10 = 23 m s$^{-1}$, therefore speed = 23 m s$^{-1}$
   (c) $2.1 \times 10^4$ J
   (d) Change in potential energy = $3.9 \times 10^4$ J, height of ramp = 49 m.
   (e) (i) 4.0 s
       (ii) 93 m

4. (a) 6.0 N
   (b) $7.1 \times 10^2$ m s$^{-2}$
   (c) 0.15 J
   (d) Friction between ball and surface.
       The spring retains some kinetic energy.
       Hysteresis within the spring.

5. (a) Coal: capital = £870 million
       operating = £2100 million
       fuel = £3400 million
       Nuclear: capital = £1400 million
       operating = £1800 million
       fuel = £260 million
   (b) High cost of decommissioning for nuclear power compared with fossil fuel.
   (c) Large quantities of highly radioactive material.
       Long time before radioactivity decays.

## Page 33

1. (a) 68 J  (b) 68 J
   (c) 46 m

2. (a) 7.3 m  (b) 2.5 MW

3. 28 m s$^{-1}$

4. (a) (i) 6.3 m s$^{-1}$
       (ii) 4.8 m s$^{-1}$

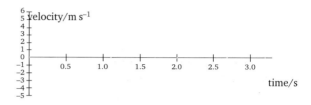

## Page 35

1. (a) $v_h$ decreasing
       $v_v$ increasing downwards or decreasing upwards
   (b) $v_h$ decreasing
       $v_v$ increasing downwards or decreasing upwards

## Page 36

1. (a) 3.5 m s$^{-2}$  (b) 12 N  (c) 22 N

3. $2.3 \times 10^3$ N

## Page 38

1. (a) In equilibrium.
   (b) Not in equilibrium; acceleration = 1.7 m s$^{-2}$ vertically upwards.
   (c) In equilibrium.
   (d) Not in equilibrium; acceleration = 2.4 m s$^{-2}$ at an angle of 45° above the vertical, to the left.

2. (a) Normal reaction = 10 N; friction = 6.0 N
   (b) Normal reaction = 19 N; friction = 16 N

3. (a) (i) 7.5 kN
       (ii) 3.50 kN
       (iii) 3.68 kN
       (iv) 6.37 kN
       (v) Combined normal reaction = 6.37 kN

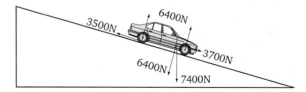

Not in equilibrium; acceleration = 0.3 ms$^{-2}$ down the slope

## Page 41

**1 (a)**

| Extension, $\Delta l/10^{-3}$ m | Time for 20 oscillations/s | | | Periodic time, $T$/s | $T^2/s^2$ |
|---|---|---|---|---|---|
| | 1st time | 2nd time | Average | | |
| 17 | 5.1 | 5.5 | 5.3 | 0.27 | 0.07 |
| 34 | 7.4 | 7.6 | 7.5 | 0.38 | 0.14 |
| 51 | 9.1 | 9.0 | 9.1 | 0.46 | 0.21 |
| 68 | 10.4 | 10.7 | 10.6 | 0.53 | 0.28 |
| 84 | 11.6 | 11.4 | 11.5 | 0.58 | 0.34 |
| 101 | 12.9 | 12.7 | 12.8 | 0.64 | 0.41 |
| 118 | 13.9 | 13.8 | 13.9 | 0.70 | 0.49 |

(b) Gradient = 4.1 $s^2$ $m^{-1}$, $g$ = 9.6 m $s^{-2}$

(c) Lower values of $\Delta l$ and $T$ are likely to be more inaccurate. Percentage error is likely to be large when measured quantity is small.

## Page 43

1 (a) 6.0 A     (b) 4.8 mA
2 (a) 240 C    (b) 0.30 C
3 (a) 12 s     (b) $1.3 \times 10^3$ s
4 (a) $1.5 \times 10^{21}$   (b) $1.9 \times 10^{18}$
5 $2.2 \times 10^{-3}$ m $s^{-1}$

## Page 44

**1**

| V | I | R |
|---|---|---|
| 12 V | 4.0 A | 3.0 Ω |
| 230 V | 18 mA | $1.3 \times 10^4$ Ω |
| 450 V | 2.5 A | 180 Ω |
| 92 V | 3.4 mA | 27 kΩ |
| 230 V | 0.25 A | 920 Ω |
| $4.8 \times 10^6$ V | 24 mA | $2.0 \times 10^8$ Ω |

## Page 46

1 (a) 24.5 Ω     (b) $4.8 \times 10^{11}$ Ω
2 $8.4 \times 10^{-3}$ Ω, $9.3 \times 10^{-4}$ Ω, $1.5 \times 10^{-3}$ Ω
3 116 m

## Page 48

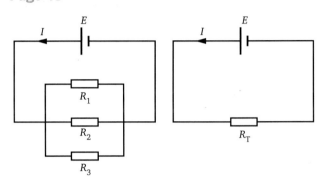

Let the current through $R_1$ be $I_1$, and so on. Charge is conserved in the circuit so $I = I_1 + I_2 + I_3$. For the three resistors:

$$I_1 = \frac{E}{R_1} \quad I_2 = \frac{E}{R_2} \quad I_3 = \frac{E}{R_3}$$

(as potential difference is the same for all components connected in parallel.)

Also $I_1 = \dfrac{E}{R_T}$

So $\dfrac{E}{R_T} = \dfrac{E}{R_1} + \dfrac{E}{R_2} + \dfrac{E}{R_3}$

i.e. $\dfrac{1}{R_T} = \dfrac{1}{R_1} + \dfrac{1}{R_2} + \dfrac{1}{R_3}$

## Page 50

1 (a) (i) 0.3 A
    (ii) 0.15 V
    (iii) 1.35 V
  (b) (i) $1.5 \times 10^{-2}$ A, $7.5 \times 10^{-3}$ V, 1.49 V
    (ii) $7.5 \times 10^{-2}$ A, $3.8 \times 10^{-2}$ V, 1.46 V
    (iii) 0.15 A, $7.5 \times 10^{-2}$ V, 1.43 V
    (iv) 1.5 A, 0.75 V, 0.75 V
    (v) 3.0 A, 1.5 V, 0 V

2 1.4 V

3 (a) $3.3 \times 10^{-3}$ Ω     (b) 13.5 V

4 Maximum current of power supply = 0.40 mA, which is below fatal level.

5  (a) $2.2 \times 10^7$ W    (b) 44%
   (c) $5.3 \times 10^3$ V    (d) 44%

## Page 51

1  (a)
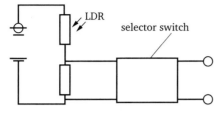

   (b)
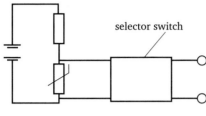

   (c)
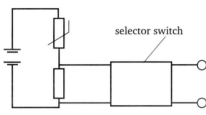

2  (a) $0.065 \, \Omega$
   (b) $0.064 \, \Omega$
   (c) see page 44

3  (a) (i) $7.5 \, \Omega$
      (ii) 0.60 A
      (iii) p.d. across internal resistor = 3.0 V
           p.d. across supply terminals = 3.0 V
      (iv) 0.60 W
   (b) resistance of parallel combination is always less than the resistance of any of the individual resistors. Therefore the resistance of the parallel combination cannot be $5.0 \, \Omega$
   (c) (i) $4.6 \times 10^{-7} \, \Omega$m
       (ii)
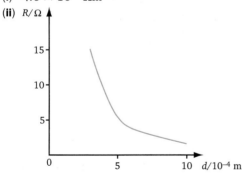

4  (a) 2.0 A
   (b) 24 W
   (c) $0.50$ m s$^{-1}$
   (d) $4.2 \times 10^{-8} \, \Omega$m

5  (a) $600 \, \Omega$
   (b) 1.0 C
   (c) See page 51
   (d) 1.5 J

## Unit 2

## Page 55

1  The time for one complete cycle of the oscillation. This is shown as $T$ on the diagram.

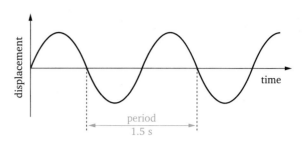

2  (a) 0.50 s
   (b) 20 ms (0.020 s)
   (c) 4.0 ms ($4.0 \times 10^{-3}$ s)
   (d) 33 μs ($3.3 \times 10^{-5}$ s)
   (e) 40 ns ($4.0 \times 10^{-8}$ s)
   (f) 0.25 ns ($2.5 \times 10^{-10}$ s)

3  (a) 4.0 Hz
   (b) $7.1 \times 10^2$ Hz; 710 Hz
   (c) $8.3 \times 10^{-3}$ Hz; 8.3 mHz
   (d) $2.0 \times 10^5$ Hz; 200 kHz; 0.20 MHz
   (e) $2.1 \times 10^7$ Hz; 21 MHz
   (f) 120 Hz

4
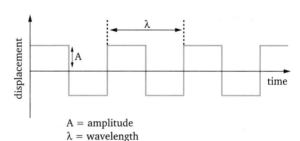

   A = amplitude
   λ = wavelength

5  For example:
   • mass oscillating on springs such as in a washing machine
   • pendulum as used in clocks
   • oscillations of bridges and buildings in strong winds

6
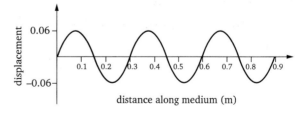

7  (a) 4.8 ms
   (b) 16 mm
   (c) 210 Hz
   (d)

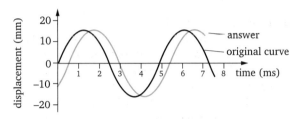

8

| Frequency | Velocity | Wavelength |
|---|---|---|
| 15 Hz | 120 m s$^{-1}$ | 8.0 m |
| 3.8 MHz | 5000 m s$^{-1}$ | 1.3 mm |
| 1.2 × 10$^{15}$ Hz | 2.0 × 10$^{8}$ m s$^{-1}$ | 0.17 mm |
| 230 Hz | 340 m s$^{-1}$ | 1.48 m |
| 11 GHz | 3.0 × 10$^{8}$ m s$^{-1}$ | 2.8 cm |

9  (a) 7.23 ms    (b) 138 (140) Hz

Page 58

1  (a) 5.9 km
   (b) Object is smaller so it reflects less energy. Wave transmitted has to travel further so energy is more dispersed.

Page 59

1  (a) 4.9 m
   (b) Reflections from fish at different depths.

2  (a) ≈340 m s$^{-1}$
   (b) ≈1.0 ms
   (c) See text for experiment 1.

3  0.28 m

Page 63

1  (a) 7.0 W m$^{-2}$    (b) 0.38 m

2  2.6 × 10$^{16}$ W

3  For example:
   (a) when reflected from the face of a large cliff
   (b) in an anechoic chamber or when incident on a heavily curtained wall
   (c) when incident on a 'flexible' solid surface such as a large glass window.

4  Along the interface between the two surfaces.

5  (a) Rotate receiver but keep it always directed towards the source. The signal goes from maximum to minimum, etc. as the alignment changes.
   (b) Ultrasound is a longitudinal wave. Polarization is a property of transverse waves.

6  (a) (i) See pages 56–57.
       (ii) See page 62.
   (b) (i) 0.015 s
       (ii) 0.0137 s
       (iii) It would have to be moved nearer to the source. If the separation is increased, the time to travel the extra distance horizontally will be less in the granite than in the direct soil route so the difference in times will increase.

7  (a) 3.0 m
   (b) Energy being spread out over a wider area or length.
       Transmitting medium absorbing energy.
   (c) Wave should show zeros shifted 0.9 m to the right.
       Amplitude of the first peak lower than it was before.
   (d) 100 Hz

8  (a) See pages 58–59.
   (b) See page 61.
   (c) 0.20 m

Page 67

1  (a) 0.016°    (b) 5.7°
   (c) 42(.5)°

2  (a) Brighter image.
   (b) Narrower central maximum but brighter.
   (c) Narrower central maximum.

3  (a) 9.8° either side of the strongest signal (total angle 19.6°).
   (b) 27° either side of the strongest signal (total angle 54°).

4  11.5 km

5  4.2 × 10$^{-5}$ light years

6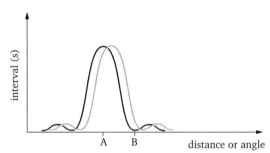

Peak of blue curve must be between A and B.

7  (a) 7.5 × 10$^{-5}$ m
   (b) The light would not be resolved. Minimum angle for resolution is increased. Need to move closer to the screen to resolve the pixels.

8  (a) 2.9 × 10$^{-6}$°
   (b) 1.2 × 10$^{-5}$°

## Page 69

1. General shapes of the graphs you should have drawn by adding corresponding displacements.

   (a)
   
   (b)

   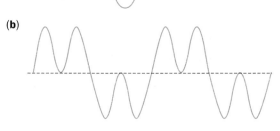

2. Your graph should be sinusoidal with amplitude 1.4 times that of one of the waves and having the same wavelength.

3. 2

## Page 74

1. (a) Maximum; path difference = $4\lambda$
   (b) Minimum; path difference = $2.5\lambda$

2. 660 nm

3. 17 μm

4. (a) 0.089°   (b) 1.3 m

5. 3.2 cm

6. 1.6 cm

7. 295 nm

8. $7.20 \times 10^{-7}$ m

9. (a) 300 m   (b) 150 m

10. (a) (i) No path difference; waves arrive in phase so constructive interference.
    (ii) Path difference is half a wavelength; waves arrive in antiphase so destructive interference.
    (b) (i) 2.7 cm
    (ii) $1.1 \times 10^{10}$ Hz

## Page 76

1. $2.2 \times 10^{-6}$ m

2. 588 lines per metre

3. (a) $2.0 \times 10^{-6}$ m   (b) 500 lines per mm

4. 0, ±15.1°, ±31.3°, ±51.3°

5. 8th

6. (a) $3.3(3) \times 10^{-6}$ m
   (b) 0.35°
   (c) 9.2 mm

7. (a) Sources with same frequency and constant phase difference.
   (b) (i) $4.0 (3.96) \times 10^{-7}$ m
   (ii) 99 nm; path difference has to be half a wavelength so thickness is a quarter of a wavelength.

8. (a) (i) See question 7a.
   (ii) The regions of destructive interference are darker so there is better contrast between light and dark areas.
   (b) (i) $1.5 \times 10^6$ lines per metre.
   (ii) The new wavelength will have maxima at smaller angles in each order. The maxima will be approximately equally spaced.

## Page 81

1. (a)

   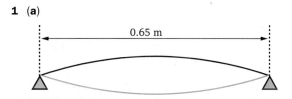

   (b) 600 Hz; 0.22 m
   (c) Four times original tension.

2. 0.32 m

3. Point of zero amplitude.

4. $2.0 (1.95) \times 10^{-3}$ kg m$^{-1}$

5.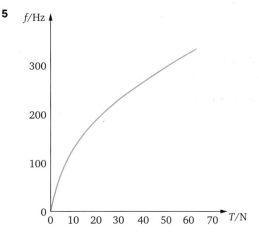

6. (a) 256 Hz
   (b) Reduce to $\frac{1}{4}$ original tension.

7. (a) 170 m s$^{-1}$
   (b) At 56 Hz, wavelength = 3 m. With the same tension and mass per unit length, the length of the string required would be 1.5 m. The string is therefore too short.
   (c) 70 Hz, 140 Hz, 210 Hz

## Page 85

1. Gamma, ultraviolet, green, red, microwaves
2. Radio waves, infrared, blue, ultraviolet, X-rays
3. See pages 82–83.
4. (a)
$$E = \frac{6.6 \times 10^{-34} c}{\lambda} = \frac{2.0 \times 10^{-25}}{\lambda}$$
   (b) (i) $4.0 \times 10^{-19}$ J
   (ii) $1.7 \times 10^{-23}$ J
5. See page 83.
6. See page 84.

## Page 88

1. (a) (i) 1620 Hz  (ii) 1400 Hz
   (b) (i) 1610 Hz  (ii) 1390 Hz
2. Using approximate formula:
   (a) 65 m s$^{-1}$  (b) 970 Hz
   Using accurate formula:
   (a) 55 m s$^{-1}$  (b) 1030 Hz
3. 41.2 kHz
4. 0.036 m s$^{-1}$
5. $6.8 \times 10^{14}$ Hz
6. (a) $9.5 \times 10^{15}$ m
   (b) 114 years
   (c) 38.5 Mpc or $1.3 \times 10^8$ light years
   (d) 6500 km s$^{-1}$
7. (a) 5800 Mpc or $1.8 \times 10^{26}$ m
   (b) 3600 Mpc or $1.1 \times 10^{26}$ m
8. (a) $6.2 \times 10^{17}$ s  (b) $3.9 \times 10^{17}$ s

## Page 90

1. A nucleus with 79 protons, 118 neutrons and 79 electrons in its atomic structure.
2. (a) $4.0 \times 10^{-25}$ kg
   (b) $5.1 \times 10^{-27}$ kg
   (c) $6.8 \times 10^{-27}$ kg
3. (a) 6  (b) 8
4. See page 89.
5.

| Symbol | $^{241}_{95}$Am | $^{220}_{86}$Rn | $^{237}_{93}$Np | $^{240}_{94}$Pu | $^{40}_{19}$K | $^{60}_{27}$Co |
|---|---|---|---|---|---|---|
| Proton number | 95 | 86 | 93 | 94 | 19 | 27 |
| Nucleon number | 241 | 220 | 237 | 240 | 40 | 60 |
| Neutron number | 146 | 134 | 144 | 146 | 21 | 33 |
| Electron number | 95 | 86 | 93 | 94 | 19 | 27 |

## Page 93

1. $^{1}_{1}$p  $^{0}_{-1}$e  $^{4}_{2}\alpha$  $^{1}_{0}$n  $^{0}_{1}\gamma$  $^{0}_{0}\nu$
2. Proton number 84; nucleon number 210
3. Proton number 84; nucleon number 215
4. (a) $^{1}_{0}$n $\Rightarrow$ $^{1}_{1}$p + $^{0}_{-1}\beta$ + $^{0}_{0}\bar{\nu}$
   (b) $^{11}_{6}$C $\Rightarrow$ $^{11}_{5}$B + $^{0}_{+1}\beta$ + $^{0}_{0}\nu$
   (c) $^{226}_{86}$Ra + $^{222}_{84}$Rn + $^{4}_{2}\alpha$
5. (a) A beta$^{-}$ particle
   (b) See page 93
   (c) $^{129}_{53}$I $\Rightarrow$ $^{129}_{54}$Xe + $^{0}_{-1}\beta$ + $^{0}_{0}\bar{\nu}$
6. (a) $1.47 \times 10^7$ m s$^{-1}$
   (b) $2.54 \times 10^5$ m s$^{-1}$
   (c) $1.27 \times 10^{-14}$ J
   (d) 1.74%

## Page 95

1. (a) $3.7 \times 10^5$
   (b) Shorter range, more atoms per unit volume so more collisions per second.
2. See page 94.
3. See page 94.
4. (a) Alpha particle slows down so there are more collisions per mm.
   (b) 5.8 cm
   (c) Graph showing higher ionization at start and shorter range. Otherwise same shape.

## Page 98

1. Half-life is ≈70 s at the start but graph data give longer half-life for lower count rates.

2.

| Source | Mg-23 | Sr-90 | U-238 | Pa-234 |
|---|---|---|---|---|
| Half-life | 12 s | $8.8 \times 10^8$ s | $4.5 \times 10^9$ years | 72 s |
| Decay constant ($s^{-1}$) | 0.058 | $7.8 \times 10^{-10}$ | $4.9 \times 10^{-18}$ | $9.6 \times 10^{-3}$ |
| Activity (Bq) | 20 | $2.0 \times 10^{11}$ | 45 | $2.0 \times 10^4$ |
| Number of radioactive nuclei | 350 | $2.5 \times 10^{20}$ | $9.3 \times 10^{18}$ | $2.1 \times 10^6$ |

3. (a) Graph showing number of radioactive atoms halving every 30 minutes.
   (b) $3.8 \times 10^{-4}\,s^{-1}$   (c) $6.1 \times 10^{16}$ Bq

4. (a) $6.5 \times 10^{-10}$ g   (b) 106 Bq
   (c) 11200 years

5. (a) (i) $0.058\,h^{-1}$
       (ii) $1.6 \times 10^{-5}\,s^{-1}$
   (b) (i) 50%
       (ii) 25%

6. (a) $3.33 \times 10^{-7}\,s^{-1}$
   (b) $2.0 \times 10^6$ Bq
   (c) (i) $\approx 4.2 \times 10^{12}$
       (ii) ≈48 days

7. (a) See page 96.
   (b) Chemically the same as other isotopes so are used in the same way by the body. Emissions from them can destroy malignant tissue by ionization.
   (c) (i) Total number of decays that have taken place up to that time.
       (ii) 8 days
       (iii) $1.0 \times 10^{-6}\,s^{-1}$
   (d) (i) $2.8 \times 10^8$
       (ii) Shape as Figure 2.91. Change y-axis label to 'number of atoms' and change the scale so that 300 becomes $300 \times 10^6$.
   (e) See page 97.

8. (a) (i) One disintegration (decay) per second.
       (ii) See page 103.
       (iii) Alpha particles are absorbed in a short range and cannot be monitored outside the body. Beta and gamma radiations have longer ranges.
   (b) (i) $2.1 \times 10^9$ Bq
       (ii) $1.5 \times 10^{13}$ atoms

## Page 102

1. (a) Maximum distance the alpha particle travels.
   (b) Gamma radiation is never completely absorbed. The intensity is reduced to half its original intensity by a particular thickness of absorber. The actual thickness depends on the absorbing material.

2. Insert different materials between source and detector. If there are α particles a thin sheet of paper reduces count rate considerably. If a few mm of aluminium is inserted the count rate should be similar to that with paper.

3. (a) See page 59.
   (b) Few γ ray photons are removed by air. At a distance $r$ from the source the photons are evenly distributed over the surface area of a sphere of radius $4\pi r^2$.
   (c) Every time the distance from the source is doubled the intensity falls to $\frac{1}{4}$ so the further away you are the lower the dose received.

## Page 103

1. (a) $70\,s^{-1}$   (b) $160\,s^{-1}$

2. See previous section.

3. (a) (i) 95
       (ii) 241
   (b) See page 94.

4. $1.6 \times 10^7\,m\,s^{-1}$

5. See previous section.

## Page 109

1.

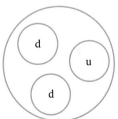

2. $\mu^- \to e^- + \nu_\mu + \bar{\nu}_e$
   Q: $+1 \to -1 + 1 - 1$
   L: $+1 \to +1 + 1 - 1$
   Both conservation rules satisfied.

   $\mu^- \to e^- + \nu_\mu + \nu_e$
   Q: $-1 \to -1 + 0 + 0$
   L: $+1 \to +1 + 1 + 1$
   Charge is conserved but not Lepton number.

3. Charge is always $+1$ or $-1$.
   e.g. an up-quark $(+\frac{2}{3})$ and an antidown-quark $(+\frac{1}{3})$
   an anti up-quark $-\frac{2}{3}$ and a down-quark $(+\frac{1}{3})$

## Page 112

**1** (a) 6Ω
(b) 0.34 V

## Page 115

**1**

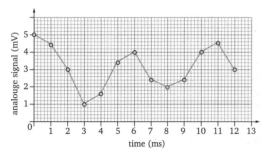

**2** (a) 37.5 mV  (b) 6.0 mA
(c) 0.23 mW  (d) ≈3.16 V
(e) ≈4.7 km

**3** (a) Set up the circuit shown below using a rotary potentiometer.

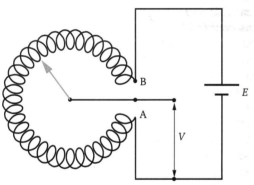

The angle turned through is:

$\frac{V}{E}$ × total angle turned through from A to B

V is 0 when the slider is at A and E when at B. The angle from A to B is typically about 270°.

(b) The separation of the windings on the potentiometer; the smallest detectable change in voltmeter reading; the supply emf.

**4** See previous section.

## Page 119

**1** 3.2 m

**2** 325.5 kHz to 334.5 kHz

**3** The intensity of each of the three primary colours, red, blue and green, at points on the screen that are one pixel apart. The image has to be scanned to determine the intensity of each colour and the receiver synchronized to the scanned image. A synchronizing signal has to be transmitted. (The eye combines the three colours to reconstitute the image at each point.)

**4** 220 kHz

## Page 121

**1** (a) 320 kbits s$^{-1}$
(b) 192 Mbits s$^{-1}$

**2** (a) About 38 000
(b) 12

**3**

| Sample time (ms) | Voltage transmitted (V) | Digital signal (uppermost bit sent first) |
|---|---|---|
| 1 | 4.0 | 0<br>0<br>0<br>1 |
| 2 | 4.0 | 0<br>0<br>0<br>1 |
| 3 | 4.0 | 0<br>0<br>0<br>1 |
| 4 | 0.5 | 1<br>0<br>0<br>0 |
| 5 | 0.5 | 1<br>0<br>0<br>0 |
| 6 | 4.0 | 0<br>0<br>0<br>1 |
| 7 | 4.0 | 0<br>0<br>0<br>1 |
| 8 | 4.0 | 0<br>0<br>0<br>1 |
| 9 | 0.5 | 1<br>0<br>0<br>0 |
| 10 | 0.5 | 1<br>0<br>0<br>0 |

# Index

Absolute uncertainty  14
Absorption experiments  100
   of waves  61
   spectra  84, 85
Acceleration  23
   due to gravity  25, 37, 39
   non-uniform  26
   uniform  24
Aerials  66, 73
Age of universe  88
Air resistance  36
Aircraft wings, lift  34
Alpha emission,  91
   effect on nucleus  92
   particle experiments  100
   speed of particle  91
Alternating current  50
AM  117
Ampere  42
Amplitude  39, 53
Analogue information  116
   instruments  110
   to digital conversion  113
Analysing  11
   with graphs  11
   without graphs  13
Antinodes  78
Antiparticles  107
Atomic structure  89
Audio information  116

Background radiation  95, 97
Band spectrum  84
Bandwidth  116
Baryon  108
   number  107
Base bandwidth  117
Bernoulli  34
Beta⁻ particle  91
   effect on nucleus  92, 93
   emission  92
   particle experiments  101
   speed of  92
Binary numbers  113
Biomass  30, 31
Bit  113
Bubble chambers  106
   use of  107
Byte  113, 119

Cable, copper  114
Cable, fibre optic  62, 114
Calculating moments  20
Carrier frequency  118

Channel bandwidth  118
   communication  56
Charge  42, 107
Charge carriers  43
Classification of particles  107
Cloud chambers  106
Coherent sources  69
Combining uncertainties  10
Communication channel  56
   systems  115, 116
Complex waves  56
Compression  57
Computer monitoring  110
Conduction, gases  43
   liquids  43
   metallic  43
Conservation rules  109
Constructive interference  68
Continuous spectrum  84
Conventional current  42
Converting analogue to digital  120
Coulomb  42
Count rate  95
Critical angle  61, 62
Current  42
   at junctions  48
   heating effect  44
   measurement sensor  111

Damping  40
Data capture, advantages  110
   logging  40
   sampling  113
   transmission  113
Decay constant  96
   equation  96
   of neutron  109
Demodulation  118
Destructive interference  68
Deuterium  90
Diffraction  64
   effects of  65
   gratings  75
   in radio transmission  65
Digital data  112
   readout  110
Digitizing information  119
Diodes  45
Diodes, rectification using  46
Distance–time graphs  23, 26
Doppler effect  86, 87
Drag  35
Drift velocity  42

# Index

Electrical energy sources 29
Electrical energy and power 49
Electricity 42
   cables, power loss 50
   generating 29
Electromagnetic radiation 82
   radiation, speed of 82
   spectrum 82, 83
   waves 57, 82, 83
Electromotive force 47, 48
Electron 42, 107
Electrons, free 42, 43
Electron-volt 91
EMF 47, 48
Emission spectra 84
Endoscope 63
Energy 49
   kinetic 33
   loss in transmission 114
   mechanical 53
   potential 32
   sources 30
   sources, nuclear 98
   stored in springs 28
Equations of motion, derivation 25
     using 26
Equilibrium 20, 21, 35, 37
Evaluating 14
Examinations 5
Experimental skills 11
Exponential changes 97
   test for 13

Falling objects, kinetic energy 33
Fibre optic cable 62, 114
   uses 62, 63
Flow laminar 34, 35
   streamline 34
   turbulent 34
Fluids, viscous 34
FM 117
Forces, adding 17
   on inclined plane 19
   lift 34
Formulae, sheet for exams 10
   to remember 9
Fraunhofer lines 84
Frequency spectrum 116
Frequency 53, 55
   fundamental 79
Friction 40
Fringe spacing 71
Fundamental frequency 79

Gamma, decay effect on nucleus 92
   radiation experiments 101
Gamma radiation 60, 62, 82, 91, 92
   inverse square law 95, 101
Geiger and Marsden 105
Geiger–Muller tube 100
Generating electricity 29

Gradients 12
Graph drawing 12
Gravitational PE 32
Gravity 25, 39

Hadrons 108
Half-life 96
   measurement 97, 98
Harmonics 79
Hubble constant 87
   Law 87
Hydro-electric power 29, 31

Industrial uses of radioactivity 102, 103
Information 110
Intensity 59
Interference 64, 68
   constructive 68
   destructive 68
   optical 71
   two-source 70
Interference fringes 71
   in communications 73
   using laser 76
   using light 72
   using radiowaves 72
   using reflections 72
   with microwaves 72
Internal resistance 48, 49
Inverse square law 59, 95
   for gamma 95, 101
Ionization 43, 94
   radiation 94
Ionosphere 61
Ions, lattice 42
Isotopes 89

Key words 8
Kinematics 23
Kinetic energy 33, 44, 53

Laser 76
Lattice ions 42, 44
Law, Hubble 87
Law, Ohm's 44
Laws of Motion, Newton's 37, 38
LDR 45, 51
Lepton number 107
Leptons 107
Lift forces 34
Light waves 57
   year 87
   monochromatic 71
Line of sight 65
Line spectrum 84
Longitudinal waves 57
Lost voltage 49
Loudspeaker design 66

Mains power 50
Mass 37, 107

Mass/spring oscillator   40
Maximum speed of vehicle   36
Measuring gravity   25, 39
Measuring sound speed   58
Mechanical energy   53
    oscillations   39
Medical uses of radioactivity   103
Mesons   108
Metallic conductors   43
    resistance   44
Mode of vibration   78
Modulation   118
Moments, calculating   20
Monochromatic light   71
Motion graphs   23
Multiplexing, time-division   121
Muon   107, 108

Neutrinos   93
    evidence for   93
Neutron   89
    decay   109
    discovery   106
Neutron–proton graph   90
Newton's Laws of Motion   37, 38
Nodes   78
Noise   117
Nuclear equations   92
Nuclear nomenclature   89
Nucleon   89
Nucleon number   89
Nucleus   89, 105
    existence of   105
Nuclides   89

Ohm   43
Ohm's Law   44
Oscillations, damping   40
    mechanical   39
    speed–time graph   24, 26
Oscillator, data logging   40
    mass/spring   40
Overtones   79

Parallel circuits   47
    data transmission   114
    resistors   48
Parsec   87
Particle decay, rules   109
    classification   107
    oscillation   54, 78
Path difference   70, 72
PE   32
Penetrating power   95
Period   39, 53
Periodic motion   53
Phase difference   55
Phase   54, 55, 68, 78
Photons   82
Pion   108
Pixels   67
Plane waves   54

Planning   5, 11
Polarization   62
Position sensor   111
Positron   92, 107
Potential difference   46, 51
    dividers   51
    energy   32
Potentiometers   51
Power distribution   30
    loss in transmission   50
Power   49
    geothermal   30, 31
    hydro-electric   29, 31
    nuclear   31
    solar   30, 31
    wave   29, 31
    wind   29, 31
Practical work   5, 6, 11
    analysing   5, 12, 13
    evaluating   5, 14
    implementing   5, 11
    planning   5, 11
Principle of moments   20
    of superposition   68
Projectile motion   27
Proton   89, 105
    discovery of   105
Proton number   89

Quarks   107, 108

Radar   57
Radiation
    activity   95
    background   95
    effect on materials   102
    detectors   100
    electromagnetic   82
    experiments   100
    penetrating power   95
    physiological effects   102
    range of   95
    safety rules   100
    uses of   102
Radio waves   57
Radioactivity   91
    randomness   96, 97
Radiocarbon dating   98
Range of hearing   116
Rarefaction   57
Recession speed   87
Recoil of nucleus   91, 93
Rectification   46
Red shift   87
Reflection   60
    partial   61
Refraction   60
    in communications   61
Remote sensing   110

Resistance  43
   air  36
   change with temperature  44
   factors affecting  44
   of filament lamp  45
   internal  48
   thermistors  45
Resistivity  44, 45
Resistors in parallel  48
   in series  48
   light dependent  45
Resolution  65, 66, 67
Resultant force  18
Revising  7

Sampling data  113
   rates  119
Scalars  17
Scale drawing  17
Sensors  111
Serial transmission  114
Series circuits  47
   resistors  48
Slit spacing  75
Solar power  30, 31
Sound synthesis  68
   waves  57
   speed of  58
Spectra, absorption  84
   emission  84
   gamma ray  84
Spectrum, band  84
   continuous  84
   line  84
Speech transmission  117
Speed  23
Speed of sound  58
Speed–time graph  24, 26
Springs  28
   energy stored in  28
Stationary waves  78
   experiments  80
Stored energy in springs  28
Stretched springs  28
Stretched string  78, 89
   experiments  80
Sun  84
Superconductivity  46
Superposition of waves  68
Synthesis of sounds  68

Tau  107, 108
Telescopes  67
Temperature, transition  46
Thermistor  45, 112
Time-division multiplexing  121
Total internal reflection  61
Transformers  50
Transition temperature  46
Transmission of electricity  30
Transmitting data  113
Transverse waves  56, 78, 80

Travelling waves  54
Tritium  90
Turbine  29
Tweeters  66
Two-source interference  70

Ultrasound  57, 61
   compared with X-rays  114
   uses  67
Uncertainties  14
Universe, Age of  88

VDU  67
Vectors  17
   resolving  18
Vehicle speeds  36
Velocity  23, 55
   resolving  19
   drift  42
VHF  65
Video information  118
Viscous fluids  34
Voltage  47

Wave front  54
   intensity  59
   packet  82
   power  29, 31
   speeds  57
Wavelength  54
Waves  53
   absorption of  61
   complex  56
   compression/rarefaction  57
   electromagnetic  57, 82, 83
   inverse square law  59
   light  57
   longitudinal  57
   mechanical  56
   particle oscillation  54
   plane  54
   properties  59
   radio  57
   sound  57
   stationary  78
   superposition of  68
   transverse  56, 78, 80
   travelling  54
Weight  37
Wind power  29, 31
Woofers  66
Work  32

X-rays  62
   compared with ultrasound  114
   uses  67

Young's experiment  71